I0866358

Designing and Creating
A COASTAL GARDEN

Alan Edmondson & Bryn Edmondson

Designing and Creating A COASTAL GARDEN

THE CROWOOD PRESS

First published in 2023 by
The Crowood Press Ltd
Ramsbury, Marlborough
Wiltshire SN8 2HR

enquiries@crowood.com
www.crowood.com

© Alan Edmondson and Bryn Edmondson 2023

All rights reserved. No part of this publication may be reproduced or transmitted in any form or by any means, electronic or mechanical, including photocopy, recording, or any information storage and retrieval system, without permission in writing from the publishers.

British Library Cataloguing-in-Publication Data
A catalogue record for this book is available from the British Library.

ISBN 978 0 71984170 5

Cover design: Sergey Tsvetkov

Acknowledgements
This book could not have been written without the support and creative inspiration of Angela Edmondson. We would like to dedicate it to her and to Fred Edmondson, much loved father and grandfather, who taught a very young Alan to love gardening and was the beginning of it all.
We would also like to thank The Crowood Press, whose patient guidance was invaluable.

Picture acknowledgements
We are very grateful to all our clients who have allowed us to work on their gardens and develop our design practice. Our particular thanks go to Ewan McGregor, Alex Finch, Mr and Mrs Bishop, Richard and Jenny Harris, and Mr and Mrs Stockwell, who allowed us to photograph their gardens for this book. We would also like to thank James Chichester of Chichester Trees and Shrubs, Neil Lucas of Knoll Gardens, and Compton Acres Gardens for allowing us to photograph.

The very talented Richard Bloom supplied us with photographs of gardens by the designers Helen Elks-Smith, Andrea Rubin, Sue Townsend Garden Design and Craig Reynolds. Photographs of the gardens designed by Paul Williams, Wendy Booth and Leslie Howell, Sarah Morgan, Jennifer Gay, Julie Toll, Rose McMonigall, Sue Adcock, Lynne Marcus, and the garden at Lip na Cloiche on the Isle of Mull were taken by the equally wonderful Marianne Majerus.

Paul D. Brock allowed us to use an amazing range of his outstanding photographs of insects and we would like to thank him, along with Barry Collet who supplied some beautiful images of butterflies and a slow worm. We are also very grateful to Keith Metcalf, a leading light in Milford-on-Sea Conservation Volunteers, for supplying us with excellent photographs of birds and the Jersey tiger moth, and Timothy Rosier for his very fine images of dragonflies.

Graphic design and typesetting by Peggy & Co. Design
Printed and bound in India by Replika Press Pvt Ltd

Contents

Introduction

A GARDEN BY THE SEA

This book is going to set out, as a step-by-step process, how you can design and create a beautiful and practical garden in a coastal setting. The seaside is an exciting and inspiring place to make a garden, but it has some particular and tricky challenges. By planning carefully and using the processes and information in this book, you can make a coastal garden that will be a life-enhancing joy for years to come.

To make beautiful, successful gardens you need to know about a lot of things. You need to know about plants, of course. They are our primary materials, providing colour, scent, and form. There is a bewildering number of species for us to choose from, each with its own contributions and its own requirements. You need to know something about the hard materials, paving stones, bricks, timber, metals, and how they are put together to make patios, retaining walls, and pergolas. You will also need a good sense of how colours and forms work together and how this living work of art will change over time. Additionally, if that was not intimidating enough, each space you are working with is never the same as any other. Each garden we work on has its own combination of sun, shade, wind, shelter, soil, and topography. It has its own difficulties and its own capabilities. If the garden is in a coastal setting, those difficulties could well seem to outweigh the capabilities. You could be forgiven for giving up and deciding to pave over the whole thing.

But a garden by the sea can be an enchanting thing. For those not lucky enough to live on the coast, for whom a day trip to the seaside is an occasional treat, there is something about the first glimpse of sun reflecting off the ocean, through a break in the trees or a gap in the hills, that awakens a sense of childish excitement. To have a garden with such a view can be to have some of that feeling every day. It could be a view that greets you as you drink your first coffee at the kitchen table looking down the garden. Or it might be a view that you must explore the garden to find, rounding a corner to suddenly be presented with a beautiful ocean view and a cosy, sheltered bench in just the right spot to admire it from. Even if your coastal garden doesn't have a sea view, there is something distinctive about coastal settings. Coastal towns and villages always have a different 'feel' to their inland counterparts; there is something in the combination of the quality of light, the sounds, the smells, something about the air itself, that never quite lets you forget the near presence of the sea.

This book will help you deal with the difficulties of establishing a coastal garden. It will help you respond to the inspiration of that coastal setting to create a garden that not only copes with the conditions but has a great sense of place, somewhere that enhances and beautifies its setting. We have worked as a father and son garden design practice, based in the town of Lymington on the south coast of England, for the past twenty years. Many of the gardens we have created over that time have been in very exposed coastal settings. This book will show you the step-by-step method that we use to design a coastal garden, and give you the practical information about plants, hard landscaping materials, and techniques that you need to make that garden a long-term success. The design process we will lead you through will be very similar to the process used by professional garden designers around the world. It is not the only way to approach making a new garden; in fact, it is not at all the way most gardens are created. Most domestic gardens have, throughout history, been produced by a trial-and-error method, tweaking and experimenting over many years. This will always be the heart of good gardening. If gardening is an art form, it is the long art, something that moves, changes, and develops over the seasons and over the years. But we believe a structured design process is a very good beginning. This is particularly true for coastal gardens, where using the wrong plants or the wrong materials in an exposed position could result in costly failure. We hope that this book will also be useful to professional garden designers, perhaps starting out in their careers or designing a coastal garden for the first time, or as a reference book for inspiration and plant or material choice. But we firmly believe that non-professional gardeners, who have the inclination to devote some time and effort to creating their garden, can follow the process set out in this book to produce a result that any professional designer would be proud of.

HOW TO USE THIS BOOK

We have arranged the chapters of this book so that you can follow a process from start to finish to design and create a garden in a coastal setting. This could be your own garden or one you are working on in a professional capacity as a garden designer. You could be starting with a largely blank canvas, that is, a plot which has little or no existing planting or hard landscaping features that you wish to retain. Or the garden may have some successful planting, a perfectly good patio, some well-established shelter belts or windbreaks, but have planted areas that need improving or need some work on the overall design feel and styling.

Whatever the scale of the work that needs to be done, the process can be followed, but smaller projects may not need every step to be completed. If you just want to work on a planting scheme for a particularly exposed bed, it would still be a good idea to survey the bed, think about the practical limitations and possibilities, the colour scheme and plant textures, any hard landscaping features that might help practically or aesthetically, and produce a drawn planting plan to work from. Of course, you could also just dip into a relevant chapter to find some ideas for solving a particular problem, such as a salt-tolerant large shrub to fill a gap in a windbreak, or some ideas for coastal themed path materials. Some of the best gardens are created in this way, slowly evolving over time, experimenting, and adapting, and this book can be your guide and companion through this process.

In Chapter 1, we will first consider what it means to have a garden in a coastal setting. What are the typical conditions that the garden will be exposed to? We will then look at some specific types of coastal plant habitats. As our design practice is based in the United Kingdom, our examples will come from there, but they will be applicable to similar environments across the world; the overall design process and many of the principles and ideas will also be widely applicable. Wherever you are based, a better understanding of the habitats surrounding a coastal garden should be an important source of inspiration for your design.

Chapters 2 to 4 cover the core design process that professional garden designers use, which you could use to produce a design for your own garden. Chapter 2 sets out a step-by-step design process, beginning with the survey and working through how to produce an outline design, construction drawings and planting plans. Depending on the scope of the project you may not need to follow every step but going through the relevant parts will be an excellent start and help you achieve successful and beautiful results. Chapter 3 goes into more detail about how to ensure that any built parts of your design, such as paving or a pergola, will be suitable for coastal environments. Styling ideas and some sample construction drawings are included to help you create an attractive coastal theme with these features. Chapter 4 explains the process of producing a planting design: the practical considerations as well as ways to work with colour, texture, and seasonal changes, to create beautiful, harmonious effects. This chapter also includes planting and maintenance techniques.

Chapter 5 is in many ways the core of the book, as it lists the plants that we recommend for use in coastal gardens. Although domestic outdoor spaces have many practical functions, such as entertaining and cooking, sunbathing, children's play, sport, and exercise, if you are going to the trouble of carefully designing that space it is likely that you have some passion for plants. For us as garden designers they are the part that excites us the most and, given a free hand, we would probably be designing almost every part of the garden as a chance to show them off to their best effect. But the coastal environment can be hard on plants and our choice is usually much more limited than it would be in an inland garden. We have worked on many coastal gardens as professional garden designers and in that time have developed a list of plants that we feel we can trust to thrive in various levels of exposure to coastal conditions. This list forms the basis of Chapter 5 and although nothing is guaranteed in gardening and the safest choice of tough, coastally adapted plants can mysteriously fail to thrive, these planting choices will help you produce schemes with the best chance of success.

How we can design and maintain our gardens in the most ecologically sustainable way and how they can attract and provide habitats for wildlife has become of increasing interest and concern to amateur and professional gardeners in recent years. In Chapter 6 we will look at some ideas for how to approach these issues in a coastal garden. We will look at how we can design our garden to harness nature rather than fight it, how to use plants (often native) to increase biodiversity, and how we can reduce our use of water in the garden. This does not need to be a worthy but limiting influence on our work: it can be a source of great interest and satisfaction, particularly when we learn to notice and appreciate the wildlife that we can attract. Chapter 6 will look in some detail at the fascinating species we may be able to encourage in our gardens. Although the wildlife we look at will be that found in the United Kingdom, the approach is applicable worldwide and should be an important part of all the gardens we create.

Finally, we have included a glossary of the more technical terms that we use and have listed some of the plant nurseries that specialise in coastal plants, and some coastal gardens that are open to the public and are well worth a visit.

Chapter 1

The Coast

COASTAL CONDITIONS

Some of the largest challenges for plants exposed to coastal conditions are salt-laden winds, sea mists or fog. The effects of these are most noticeable within about 300m (980 feet) of the seashore but salt scorch on leaves has been observed as far as 10km (6 miles) inland. The effect on hard landscaping materials can be observed even further inland with, for example, coastal atmospheric corrosion a problem for the use of architectural metal within about 900m (2,950 feet) of the ocean and observable up to about 50–80km (30–50 miles) from the sea. Salt can accumulate on plants and in the soil, pulling moisture from the leaves and roots. Plants that are native to areas prone to salt-laden winds have evolved adaptations to deal with salt stress and we will look at those in more detail in Chapter 5. However, salt-laden winds may not be a wholly negative environmental factor for plants. There is some suggestion that they may have an anti-fungal effect, which is beneficial.

As well as carrying salt, the wind in most coastal areas tends to be stronger than that in low-lying inland areas due to the sea surface producing less friction. In the United Kingdom the prevailing winds are west to south-westerly and coastal areas are more exposed in the north and west of the country. Some air movement is beneficial for plants, tending to reduce humidity and so discourage fungal diseases, but high winds increase the amount of transpiration from leaf surfaces, increasing the stress from water loss. Strong, gusting winds can also cause physical damage to the plant's structure, rocking it, loosening roots, and breaking off leaves and branches. The effect of this wind pruning can be seen in the typically low-growing shrubs in exposed coastal regions, shaped by the strength and direction of the prevailing winds coming off the sea.

The wind in coastal areas often exhibits a typical pattern across the day. Water has a higher specific heat capacity than soil and rock and so the sea takes longer to heat up and cool down than the land.

These oaks, on a path through coastal salt marshes, have been shaped by the prevailing wind.

Tree mallow, *Malva arborea*, is vulnerable to severe frost and so, in the United Kingdom, in the wild it is almost always found in mild microclimates within 100m of the sea. Here it is growing on a clifftop overlooking the Isle of Wight.

The yellow horned poppy (*Glaucium flavum*), native to much of the coastline of England and Wales, has a very long tap root to help its stability in grainy, sandy soils.

As the sun warms the land during the day, the air above the land also warms and rises. The air over the sea remains cooler and the pressure difference causes the air to blow in from the sea onto the land. This causes the typical onshore, or sea breeze – a cool wind blowing in from the sea during the day. At night the reverse can happen: the sea cools more slowly than the land and so the cooler, higher pressure air over the land blows out toward the lower pressure over the sea creating an offshore, or land breeze.

We will look at soil types in more detail in the next section, but sandy soils are typical. These tend be well drained, quick to warm and not subject to compaction, factors which can encourage good growth and root penetration. But the lower capacity to hold on to water and the increased levels of evaporation from the surface can be challenging in hot, dry summers. They also tend to be lower in organic matter and so are likely to need increased levels of good compost and other soil additives if you want to grow plants that are not adapted to these soils. Grainy, sandy soil also means that the roots are not as strongly held and so are more vulnerable to wind rock. Plants adapted to these conditions may have evolved deeply penetrating tap roots to deal with this effect.

Whilst many of these environmental factors are challenging for plants and so limit our choices as gardeners, making designing and establishing a garden a little more difficult, there are distinct attractions and advantages. We often associate the coast with bright sunny days and although this impression may be partly due to association with summer holidays on the beach, there is certainly a particular quality of light on clear days as the sun shines in a wide, unobstructed sky and reflects from sand and sea. It is also certainly true that coastal regions tend to have relatively milder climates. As previously mentioned, the sea takes longer to warm and longer to cool down than the land. The effect of this is observable across the seasons with coastal summers being cooler than inland summers and coastal winters warmer and less prone to frost. In the next section we will look at some of the habitats and types of environment to be found around the coast of the United Kingdom, and how we might take inspiration from their beauty and variation.

COASTAL HABITATS

Although there are some typical environmental effects we tend to find in coastal areas, there is, of course, a great deal of variation in coastal conditions and habitats. When you come to design a garden, wherever it is situated, it is a good idea to have an understanding of its context. At a very practical level it is vital to know your soil and climate but beyond that, some understanding of the plant habitats around you, the geology, and even the history of the use of the land, can feed into your design, inspiring you and setting your design firmly within the land around it.

There is, of course, a strong tradition in gardening, perhaps particularly in the British gardening tradition, of bringing in plants and design influences from across the world. This is one of the fascinating aspects of creating a garden and should be celebrated but there has been a strong recent trend for garden design to respond closely to its surroundings, something that Japanese garden design has long understood. This absolutely does not mean you need to limit yourself to plants that are local to you (although, if you are planting a wildflower meadow, there are good ecological reasons for doing that if you can). When we are designing planting schemes for coastal gardens in particular, the extended choices we get from borrowing plants from coastal environments around the world are very useful. But including at least some plants that are local to you, cultivars of those plants, stone or locally sourced landscaping materials, or any other local inspiration you can find, can very much help produce a harmonious garden design. This is why, even if your drawing skills are not great, taking a walk with your sketchbook can be very helpful in the early stages of producing a design. Photographing your locality is also an option but the act of sketching, even badly, is a good aid to close observation.

This section will take a broad look at areas of the coastline of the United Kingdom, their habitats and other features that may be important practically or as sources of inspiration for your garden design. Many of the habitats discussed will be common around the

Whatever your level of drawing ability, some time spent walking around and sketching details of the local habitat will be a very useful source of inspiration for your design.

United Kingdom and around the world; others are more localised. The habitats vary in the amount of exposure to tides: some are regularly covered, others only during exceptionally high tides, and some are beyond the extreme high tide mark but still affected by salt winds and other coastal climatic features. Those habitats within the typical intertidal zone are unlikely to provide much inspiration for our planting but if your garden is close to such a habitat, it is worth having some understanding and appreciation of it so you can perhaps design a sympathetic transition from your garden to the landscape beyond. Coastal habitats are also divided into those that are eroding, producing features such as cliffs and caves, and those that are accreting, where the sediment produced by erosion is deposited to produce landforms such as mudflats or saltmarshes.

There are several classification schemes commonly used to describe habitats. We have based this list on the UK Habitat Classification, UKHab (further details of which can be found at ukhab.org), and we have included the UKHab codes and titles in italics within our descriptions. This is not an exhaustive list of the habitats to be found along the British coastline but is intended as a list of those of most interest to us as garden designers. We have listed some of the plants and wildlife that are typical to these habitats which could be used in a planting scheme to reflect some of that habitat in your garden. Further details about how to do this can be found in Chapters 5 and 6. Some of these habitats are under threat and plants should never be taken from the wild. Care should also be taken that you are not planting invasive species, not native to these habitats, which might escape your garden and establish themselves in the wild. But some native planting, particularly using plants that are typical of habitats surrounding your garden, can be very beneficial to wildlife and can create a garden design that is harmonious with its landscape.

This intertidal zone is unlikely to provide any direct inspiration for garden planting, but a good garden design might create a tasteful transition to a view of such a habitat.

Coastal vegetated shingle

UKHab: ***s3b Coastal Vegetated Shingle***

Shingle beaches, formed from sediment with particles of 2–200mm (1⁄16 to 8 inches) in diameter, are a reasonably common feature along the British coast but unusual outside of north-western Europe, Japan and New Zealand. The high energy waves that are required to deposit sediment of this size mean that they do not usually support much plant life but on certain stable spits or similar landforms vegetation can get established and some of those plant communities are unique to shingle. If your garden is near to a shingle coast that doesn't support vegetation, this section could still apply to you and be used to help you design a garden that will work well in this setting.

The largest site in the United Kingdom is to be found at Dungeness, a headland on the coast of Kent which is of particular interest to coastal gardeners as it is the site of Derek Jarman's famous garden at Prospect Cottage. This garden is a fantastic example of how to create a garden in sympathy with its surrounding habitat and if your garden is near a shingle fringed coastline, you could take much inspiration from it. His creative use of driftwood, fishing floats and other found objects from the beach would certainly work well in such a garden. Settling these within a planted area covered in a mix of rocks, shingle and finer gravels would bring the surroundings into the garden. Jarman's garden is open to the shingle expanse around it: 'My garden's boundaries are the horizon,' as he put it. Such an area need not be the whole of your garden design; some parts could still be more sheltered and have a different feel.

Derek Jarman's clever use of native planting such as gorse (*Ulex europaeus*) and sea kale (*Crambe maritima*) makes this garden at Prospect Cottage very much part of the surrounding Coastal Vegetated Shingle.

A shingle planted area that is exposed to salt-laden winds would need a careful choice of plants but inspiration can be taken from the plants typical in this habitat. In the wild, these habitats are sometimes bordered by shrubs, such as gorse (*Ulex europaeus*) and blackthorn (*Prunus spinosa*), and stunted trees (*see* Chapter 5 for some suggested trees for your planting plan). These could be used to frame your view and create other more sheltered areas. Typical plants of the shingle that could be used in your planting scheme include sea kale (*Crambe maritima*), the yellow horned poppy (*Glaucium flavum*), sea campion (*Silene maritima*) and sea beet (*Beta vulgaris* subsp. *maritima*).

Insect species that are associated with this habitat include the wall brown butterfly (*Lasiommata megera*), brown-banded carder bee (*Bombus humilis*), the large garden bumblebee (*Bombus ruderatus*), shrill carder bee (*Bombus sylvarum*), the cuckoo bee (*Nomada ferruginata*), the bugs *Rhopalus rufus* and *Monosynamma maritima*, the leafhopper *Aphrodes duffieldi*, the dark guest ant (*Anergates atratulus*), July belle (*Scotopteryx luridata*) and the hummingbird hawk-moth (*Macroglossum stellatarum*). We have included a planting scheme (Moths and a Shingle Beach) in Chapter 6 which could work very well in a wildlife-friendly garden design inspired by this habitat.

Beaches, sandbanks and mudflats

UKHab: ***t2 Littoral Sediment***

This widespread habitat category includes the sandy beach, many people's first thought when they think of the seaside. Beaches tend to form in more exposed locations, such as open coasts and bays, where the wave action prevents more fertile silts from accumulating. These are somewhat more common in the northern and western parts of the United Kingdom. Mudflats occur in more sheltered areas, such as estuaries and inlets, and can be found all around the British coastline.

These habitats are intertidal or otherwise regularly covered by salt water, so they do not support vegetation other than some microalgae on mudflats and dwarf eelgrass in sheltered areas. However, the sediments that make up mudflats are very fertile and support dense populations of invertebrates such as polychaete worms and bivalves. These invertebrates are an important source of food for wading birds and wildfowl.

Although there are no native plants from this habitat that could be used in a planting design, we could take inspiration from some of the planting that might be found on the fringes of these areas. A scheme based on ornamental grasses would work well to frame a view of a beach or mudflat. A sandy beach could be mirrored in your design with sympathetic hard landscaping materials such

Although this mudflat at Keyhaven in Hampshire does not support a varied plant community, it is a fertile, rich food source for wading birds and wildfowl.

The patterns made by the channels of water that typically run through a saltmarsh could be an inspiration for a serpentine layout in your garden design.

The use of ornamental grasses in this garden, created by Elks-Smith Garden Design, creates a harmonious connection with the view of the natural habitat beyond.

as sandstone paving, or timber structures inspired by groynes or driftwood. A sandy path edged with low-growing native coastal planting such as beach aster (*Erigeron glaucus*) could lead the eye very well to an end-of-the-garden view of this habitat type. Such a path would probably be best achieved using a self-binding gravel such as hoggin, or perhaps crushed shells. Going out with your sketchbook or researching how local artists have interpreted this habitat could be a source of inspiration. The way channels of water meander through mudflats could be the starting point for a curvilinear themed design for planting and paths.

Your planting design could also be influenced by closely related habitats even if they themselves are not present. For instance, planting inspired by the native vegetation of the saltmarsh would work well in a garden overlooking mudflats, and machair or sand dune vegetation could work with a view of a sandy beach.

Saltmarsh

UKHab: *t2a Coastal Saltmarsh*

When mudflat sediments accumulate sufficiently to rise above regular tidal inundation, salt-tolerant plant species can begin to get established. This stabilises the area and increases the amount of sediment that is retained, raising the ground further and allowing more plant species to colonise. When this happens, a saltmarsh habitat can be created. The plants to be found vary from the low-lying edges of the mudflat, which are often covered by high tidal waters and can only support the most salt tolerant of plants, to higher areas that are only covered by extreme tides and can support a much wider variety of plants. Saltmarshes are to be found all around the coastline of the United Kingdom.

Extensive complexes are found around major inlets and estuaries in lowland areas. Smaller, more isolated saltmarshes are associated with smaller estuaries and sheltered positions at the head of sea lochs in more upland areas.

Saltmarshes support many species of invertebrate and are an important source of food and roosting for breeding, wintering, and migrating birds, particularly wildfowl and waders. The plant species present in this habitat vary quite a bit across the country. Scottish saltmarshes for instance tend not to have the abundance of sea purslane (*Atriplex portulacoides*) and sea lavenders (*Limonium* spp.) that are characteristic of saltmarshes south of the Solway Firth. So if you are keen to reflect a local saltmarsh habitat in your planting scheme you should investigate which plants are particular to your area. However, some broadly typical saltmarsh plants that could work well in a garden include sea beet (*Beta vulgaris* subsp. *maritima*), marsh mallow (*Althea officinalis*), sea plantain (*Plantago maritima*), thrift (*Armeria maritima*), and sea aster (*Aster tripolium*).

Narrow creeks are a typical feature of saltmarshes, where the tidal waters flow in and out between higher areas of sediment stabilised by plants. These could be an interesting source of inspiration for a design feature consisting perhaps of an actual stream meandering through your planting or a more abstracted representation, such as branching paths or dry 'streams' of sandstone pebbles.

Machair

UKHab: ***26 Machair***

One of the rarest habitats in Europe, this type of low-lying coastal grassland is only found in North and West Scotland and Northern and Western Ireland. Machair is formed by the action of very strong onshore winds blowing across sandy beaches and eroding sand dunes. Calcareous shell sand and seaweed build up over impermeable bedrock creating a flat or gently undulating habitat which is sustained by traditional farming techniques (light grazing, hay cutting and small-scale food production) and is very rich in flowering plant species.

Field scabious (*Knautia arvensis*) is a prominent feature of this machair in Connemara, Ireland.

A planting scheme inspired by the machair might include bird's foot trefoil (*Lotus corniculatus*), lady's bedstraw (*Galium verum*), wild thyme (*Thymus praecox*), selfheal (*Prunella vulgaris*), red clover (*Trifolium pratense*), and common harebell (*Campanula rotundifolia*). Not all these plants are recommended for a site very exposed to salt winds, so this is an effect that would be better created in a more sheltered part of a coastal garden.

Machair is a flower-rich habitat and so an important source of nectar for bees including the scarce red shanked carder bee (*Bombus ruderarius*) and moss carder bee (*Bombus muscorum*) and the very rare great yellow bumblebee (*Bombus distinguendus*) and northern colletes bee (*Colletes floralis*).

Sand dune

UKHab: ***s3a Coastal Sand Dunes***

Sand dunes can be found along much of the United Kingdom coastline and support a surprising range of plants and wildlife for such an exposed habitat. They are formed when sand, blown inland from the beach, accumulates because of an obstacle such as a plant, or because of rising ground. The root systems of grasses like sand couch (*Elymus farctus*) and marram grass (*Ammophila arenaria*) bind the sand together and stabilise the dune as it develops. These dunes are often part of a successive system from beach edge vegetation, through embryonic and mobile dunes, to fixed dune grassland. Slacks, the low-lying areas between the dunes can resemble dry or wet heathland habitats, but much of the habitat is exposed, dry, unstable, and unfertile. Consequently, the plant life tends to be very adapted to coastal conditions.

A garden set among or adjoining a system of sand dunes is a challenge but an exciting one, and one that would respond very well to planting that looked to the native planting surrounding it. Closer to the property you might use shelter belt planting or hard landscaping to create more sheltered planting areas. These could then be reduced towards the boundaries of the garden, where the planting would more and more mimic the surrounding natural planting. If the situation allows one or the more of the boundaries to be very open, perhaps just using some simple chestnut pale fencing, then the garden can really become part of the landscape in a very pleasing way. The hard landscaping could reflect structures

Well-established planting on sand dunes at Formby, Lancashire. The pine woodlands beyond are home to one of Great Britain's few surviving colonies of red squirrel (*Sciurus vulgaris*).

These soft cliffs at Barton-on-Sea in Hampshire are typically less steep and support a wide variety of plants.

found amongst the dunes or on the beach. Boardwalk paths would certainly work, and perhaps low retaining walls of timber, reminiscent of piers or groynes, to provide sheltered areas and to stabilise some dune-like sloping planting areas.

Plants you might use include sea holly (*Eryngium maritimum*), hound's tongue (*Cynoglossum officinale*), biting stonecrop (*Sedum acre*), and sea beet (*Beta maritima*). Dune-dwelling invertebrates that you might attract include the grayling butterfly (*Hipparchia semele*), the sand wasp (*Ammophila sabulosa*), the rare dune tiger beetle (*Cicindela maritima*) and the sandbowl snail (*Catinella arenaria*).

Rocky shore, cliffs and steep slopes

UKHab: *t1 Littoral Rock*
UKHab: *s2a6 Soft Rock Sea Cliffs*
UKHab: *s2a5 Vegetated Sea Cliffs*

These habitats, produced by shoreline wind or water erosion, vary greatly in the vegetation they support depending on how exposed they are to wind, salt spray, and tides. Steep slopes or vertical hard cliffs are widely present around the coastline of Britain and Ireland, particularly in more exposed areas. They tend to be formed from rocks that are more resistant to erosion such as granite, sandstone and limestone but can also form from softer rocks such as chalk. A large proportion of Europe's chalk cliffs are to be found in England. Rocky shores are often formed at the base of these steep slopes or cliffs and consist of boulders of various sizes and shapes, and rocky ledges. Soft cliffs, formed from the erosion of less resistant material such as shales or boulder clay, are not as widespread as hard cliffs, occurring mainly along the east and central south coast of England, in Cardigan Bay and along the coast of North Wales. The softer nature of these cliffs forms less-steep slopes and a more complex habitat supporting a wider variety of plant and invertebrate species.

Widely varying though these habitats are, we have grouped them together here as they suggest some very similar design ideas. Incorporating influences from these landscapes into your garden design is likely to involve using rocks and slopes – either naturally occurring or created during the landscaping. This could be a quite modest rock wall feature, a coastal garden version of the traditional rockery, or something involving greater changes of level and larger rocks. A garden that slopes away down to cliffs or a rocky shore might be designed to start nearer the house with gabion walls as a more formal echo of the habitat, creating some sheltered spots for a wider range of planting. Then, towards the end of the garden, a more natural effect could dominate the design with boulders, natural rock walls, and native planting.

The plants associated with these habitats vary very widely depending on the amount of exposure. The exposure will be greatest on features such as headlands and along the southwest and northern coasts of the United Kingdom where high waves and strong winds mean that salt spray is delivered higher up the cliffs and slopes. Rocky shores are primarily an intertidal habitat with few plants of use to us in our planting schemes. But plants such as sea lavender (*Limonium corniculatus* and *Limonium binervosum* are useful species for planting schemes) and thrift (*Armeria maritima*) can be found along the upper edge of the splash zone. These species will also be found on the more sheltered ledges of hard cliffs, the slopes of soft cliffs and on clifftop grassland along with sea beet (*Beta vulgaris* subsp. *maritima*), sea plantain (*Plantago maritima*), buck's horn plantain (*Plantago coronopus*), ribwort plantain (*Plantago lanceolata*), wild carrot (*Daucus carota*), bird's foot trefoil (*Lotus corniculatus*) and restharrow (*Ononis repens*). Invertebrates in these habitats that might be attracted to your garden if they are local to you include moths such as thrift clearwing (*Pyropteron muscaeformis*) and Barrett's marbled coronet (*Hadena luteago*), and the snail *Ponentina subvirescens*. The microhabitats of soft cliffs are particularly rich in rare invertebrates including the ground beetle *Cicindela germanica* and the rare Glanville fritillary butterfly (*Melitaea cinxia*).

COASTAL AREAS OF THE UNITED KINGDOM

The coastline of the United Kingdom consists of a remarkable variety of landscapes, habitats, and designated areas of natural beauty. It measures about 12,429 kilometres (7,723 miles) of which about 60 per cent is in Scotland and the offshore islands. Counterintuitively though, this measurement, as of all coastlines, cannot be well defined. This 'coastline paradox' means that the length of a coastline depends on the method of measurement and the amount of detail included. The fractal nature of the shape of the coastline means that any increase in the accuracy of the measurement just measures smaller and smaller features and does not increase the certainty of the measurement.

Map of the United Kingdom.

TABLE 1.1 COASTAL AREAS OF THE UK AND THEIR HABITATS

Area	Habitats
West Coast of Scotland	• Coastal vegetated shingle (south-west) • Beaches, sandbanks and mudflats • Machair • Sand dune • Rocky shore, cliffs and steep slopes (particularly hard cliffs, with soft cliffs present in Skye)
East Coast of Scotland	• Coastal vegetated shingle (north-east) • Beaches, sandbanks and mudflats • Saltmarsh (isolated and small scale) • Machair (Northern Isles) • Sand dune (excellent examples of which can be found in the Forvie National Nature Reserve, Aberdeenshire) • Rocky shore, cliffs and steep slopes (soft cliffs are particularly to be found on the coast of Fife)
Northern Ireland	• Coastal vegetated shingle (a few sites such as Ballyquintin in County Down) • Beaches, sandbanks and mudflats • Saltmarsh (a few small sites) • Machair • Sand dune (for example Portstewart Strand) • Rocky shore, cliffs and steep slopes (particularly hard cliffs with some soft cliffs along the coast of Antrim)
North East	• Beaches, sandbanks and mudflats • Sand dune • Rocky shore, cliffs and steep slopes
North West	• Coastal vegetated shingle • Beaches, sandbanks and mudflats • Saltmarsh (Liverpool Bay and Solway Firth) • Sand dune • Rocky shore, cliffs and steep slopes (particularly hard cliffs)
Yorkshire and Humberside	• Beaches, sandbanks and mudflats • Saltmarsh • Sand dune • Rocky shore, cliffs and steep slopes
Wales	• Coastal vegetated shingle (some small sites) • Beaches, sandbanks and mudflats • Saltmarsh • Sand dune (for example Kenfig Reserve, Glamorgan) • Rocky shore, cliffs and steep slopes (particularly hard cliffs in the south-west and soft cliffs in Cardigan Bay and the north-west)
East of England	• Coastal vegetated shingle (East Anglia) • Beaches, sandbanks and mudflats • Saltmarsh (The Wash) • Sand dune (for example Holkham Reserve, Norfolk)
Thames Estuary	• Beaches, sandbanks and mudflats • Saltmarsh
South East	• Coastal vegetated shingle (Dungeness is by far the biggest in the United Kingdom) • Beaches, sandbanks and mudflats • Saltmarsh • Sand dune (for example Sandwich and Pegwell Bay, Kent) • Rocky shore, cliffs and steep slopes (soft cliffs and hard chalk cliffs)
South Coast	• Coastal vegetated shingle (for example Lepe, Hampshire) • Beaches, sandbanks and mudflats • Saltmarsh (excellent examples are to be found around The Solent) • Sand dune (for example East Head, Sussex) • Rocky shore, cliffs and steep slopes (soft cliffs and hard chalk cliffs)
South West	• Beaches, sandbanks and mudflats • Saltmarsh (Severn Estuary) • Sand dune (for example Braunton Barrows, North Devon) • Rocky shore, cliffs and steep slopes (particularly hard cliffs)

Planting Section A
Scale 1:100
Project:
Client:
Bowercot
Bowercot
Garden
Design

Chapter 2

The Design Process

WHY DESIGN?

The vast majority of beautiful, practical and successful gardens did not start with a drawing. They have been the work of amateur gardeners who, over many years, experimented, tweaked and dreamt up new ideas whilst idling in the garden on a sunny afternoon. Gardening is the long art. It requires time and patience. It is like the creation of a sculpture that is never quite finished and is made from ever-changing materials. If you have the time and patience, it is the best way to make a garden and this book will help you develop the knowledge and skills to make it happen.

But there is another approach. As professional garden designers we work with clients who want a beautiful and practical outdoor space but don't have the time or expertise to let it slowly evolve over many years of work. They may be happy to spend some of their leisure time working on the garden but still want to be sure that they start with something that has the best chance of success. Design drawings are the way we communicate with these clients, checking that we have understood their needs, tastes and budget. This chapter will teach you how to produce those drawings and how to create the best starting point possible for a coastal garden. Most courses in professional garden design will teach a process very similar to the one laid out here. If you are taking one of these courses or are now working in the profession this chapter will help you apply that process to coastal gardens in particular.

Garden designers will usually use a step-by-step process to take them from initial ideas as sketches and mood boards, to detailed scale drawings that are used to build and plant the finished garden.

Whilst this process is more typical of a professional approach, these drawings are also a good first step for the amateur gardener who has perhaps just acquired a garden by the sea and is wondering where to start. If you work through this design process, it will help you understand what you want from your garden and what is possible. It will help you to come up with exciting ideas that you might not otherwise have considered. It will help you avoid wasting money on plants or features that don't work. And, if you need to employ landscapers at any stage, it will help you communicate your requirements to them.

COASTAL DESIGN CONSIDERATIONS

If you want to improve or create a new garden exposed to coastal conditions, there is perhaps an even greater argument for starting with a design drawing. It will help you plan how to deal with your greatest challenge, the salt wind. Even with careful planning you will lose more plants than if your garden was further inland. If you don't plan at least to some extent how you are going to deal with the wind you will lose many more and waste a good deal of money and effort.

With a plan you have the opportunity to think about how you will deal with the salt wind. You can decide where and how you will create shelter, either with carefully placed plantings of tough shrubs, hedging or trees that will filter and deflect the wind, or with built elements such as walls, fences, toughened glass screens, or sunken seating areas. Where these measures are not possible or desirable, you can make sure that your planned planting will be adapted to deal with the full force of unsheltered coastal conditions. Other chapters in this book will cover in detail how you can use hard landscaping features and planting in a coastal garden. This chapter will cover how to get those ideas and solutions onto a plan that can be used to build and plant the garden.

Protected by its terraced lower position, walls and fencing, this garden by Sue Townsend Garden Design achieves fantastically colourful and varied planting whilst maintaining its sea view.

A NOTE ON SOFTWARE AND PRINTERS

We will assume throughout this chapter that you are using some sort of CAD software. The software used by professional garden designers can be rather expensive, but the cost is often justified by the improved workflow. There are many options however for cheaper or even free CAD and vector drawing software that could work perfectly well. There will be a learning curve for all options and learning to use them is beyond the scope of this book. However, as well as paid courses, there are a lot of useful tutorials available online.

If you don't want to use software, perhaps because you are designing your own garden and do not feel that the cost and learning curve are justified, it is perfectly possible to use pen, pencil and paper. Your initial survey would be carefully drawn onto a sheet of A3 or A1. This can then be taped to a drawing board and subsequent stages put on sheets of tracing paper taped over the initial survey. A final design can then be drawn on a sheet of tracing paper and photocopied or scanned.

In our professional practice, we almost always work with surveys and plans at a scale of 1:100 or 1:200 printed on A3 paper. If you are a professional garden designer, you will probably want to invest in a printer that can print at A3 size. If you often work on very large gardens, you may find you need an A1 plotter, but these are quite expensive. Also, A1 plans are a bit cumbersome to work with on a windy site. On the few occasions that we find A1 size is necessary, we use the local print shop.

The design process outlined here will also help you think about how you are going to respond creatively to a coastal setting. Too many gardens suffer from ill-thought-out styling details. There are fantastic ways to reflect your coastal surroundings, but we see perhaps too many things like lobster pots or driftwood, added as an impulsive afterthought. These things can be used successfully but if you take a little time to engage your imagination and be clear about what you are trying to achieve, the end effect will be so much better.

THE SIX-STAGE PROCESS

What we will set out here is a detailed method for producing all the drawings that we would typically create as professional garden designers. It is a fairly standard process; most garden designers would produce a very similar selection of drawings. If you are not a professional and are designing your own garden, you might not need to create all of the following drawings. An outline drawing and a planting plan might be enough to work from and details of the construction of patios or other hard landscaping features could be worked out as you go or with some help from a professional landscaper. Planning for changes of level in a garden is very important but is not easy to measure accurately. If you are just working on your own garden, you might get away with just having an idea about how the levels change and work out exactly how to deal with it as the work progresses. But the full process with all the drawing produced will help anticipate problems, avoid wasting money on ordering the wrong materials or quantity of material, and allow you to communicate very clearly your requirements to a landscaper, should you employ one.

1. The brief

A design brief usually refers to something given by a client to a professional designer but is also something which you should produce for yourself when you start thinking about a design for your own garden. It is a series of questions to ask yourself or your client to find out firstly the practical considerations of what the garden will be used for and any problems that need addressing, then some more aesthetic questions about the style and effects that you or your client are looking for. It is usually the first meeting we have with a client, conducted before we have produced a quotation for how much we will charge for the design. It is usually not something that clients are charged for but if you have very impressive credentials or are just very confident, you might get away with charging for your time. Certainly, I would consider charging if there was a lot of travel involved in getting to the client.

GARDEN DESIGN BRIEF

1. **Practical needs**
   a. Family members/residents – ages? What do they want to use the garden for?
   b. Patio – how many people should be accommodated? Bistro table for morning coffee for two or a dining table for up to ten or more guests? Sun loungers? Modular furniture with a fire pit to seat six for evening drinks?
   c. Outdoor cooking – space for a barbecue? A built-in barbecue and pizza oven? Raised beds by the cooking area for herbs or vegetables?
   d. Other sitting areas – a sheltered spot that catches the evening sun, has a great view and needs an arbour bench? A shaded pergola area for comfortable outdoor dining? A sunny terrace with fragrant planting for lounging?
   e. Recreation – enough lawn for the children to play football? A swimming pool? Space for badminton in the summer? A wooded area for the children to make camps and dens? Hard standing for dinghy storage? Outdoor cinema?
   f. Productive – vegetables? Herbs? Fruit trees? Soft fruit? Cut flowers?
   g. Wildlife – habitat creation and food plants for insects? Encourage birds to visit? A meadow of native wildflowers?
   h. Water – formal water feature? Wildlife pond?
   i. Maintenance – how much garden maintenance is acceptable? Daily? Weekend only? Just keep the lawn cut and a bit of pruning a few times a year? A no-maintenance garden doesn't exist and coastal gardens tend to be a bit more demanding in upkeep, but it can be kept low.

2. **Problems to be addressed**
   a. Screening – ugly adjoining building that needs hiding? Windows of nearby house that need screening? Windy areas that need shelter?
   b. Levels and slopes – does the lawn slope too much to be usable? Does the garden need to be terraced? Does the garden slope towards the house creating potential drainage problems?
   c. Drainage – are there areas of the garden that are too wet to be lawn? Do they need drainage to be installed or could they become bog gardens or a pond area?
   d. Existing planting – are there trees that have become too old, unsightly or unsafe? An area of wild rhododendron that could be removed or drastically cut back to regain the space? It is important to be unsentimental here and not keep things that have outlived their usefulness but at the same time to be aware of Tree Protection Orders, maintaining some maturity and not losing valuable screening or wildlife habitat.
   e. Wildlife – are rabbits or deer a problem?

3. **Style**
   a. Is a particular garden style desired? Minimalist and modern? Cottage garden? Prairie planting? Chic and restrained? Tropical foliage and hot colours? Mediterranean courtyard? Photographs of gardens that you or your client like can be helpful, as can a mood board of inspirational images or colour swatches.
   b. What style is the house? A house with a very modern style of architecture is going to suit a more modern style of garden with clean lines, drifts of perennials and architectural shrubs. A contrast can be exciting though: a beautiful nineteenth-century cottage with some very modern planting can create a great effect.
   c. Any particular colours or combinations of colour, either to use or absolutely avoid?
   d. Lots of architectural foliage or lots of colourful flowers? Restrained or exuberant?

4. **Budget**

If you are designing your own garden, you should have a pretty good idea of what you want to spend. If you are working for a client, it can be trickier. Some clients are wary of giving a budget in case you just work up to the budget unnecessarily. I tend to talk about ballpark ranges of budget in the initial meeting, whilst being careful not to promise anything too specific at this early stage. For my practice, in 2022, for a full garden redesign, build and plant, nothing much costs less than £8,000; £15,000–£30,000 for the usual mid-range, mid-sized gardens; £40,000–£70,000 for bigger properties. But different designers may find they mostly work on a much smaller or much grander scale. The ballpark price ranges you give to clients may come as a surprise. Garden design makeover TV shows which typically quote budgets that include only material costs and ignore labour, mark-up and design fees have not helped. But a well-designed, professionally built and planted garden can add to a property's resale value and, more importantly, can be one of the most enjoyed and life-enriching parts of it.

A mood board is a useful tool for collecting and presenting your ideas about materials, planting and colour schemes.

For a professional, how well you listen to your client during your initial meeting will govern how well the whole design process will go. Listening is vital. You must put aside your own initial ideas of what the garden needs and how it should be styled and listen out for every clue from the client about what is going to make them happy. Sometimes they might not have formed a very clear idea yet. You need the right questions to draw them out. The right language can also help. Do they use visual language like 'I see a patio going here' or 'I don't want bright, dazzling colours'? Do they describe things in terms of feeling, like 'I'm not happy with the way these trees loom over the garden'? Or do they talk more about practical situations like 'I want to sit here with a glass of wine in the evening' or 'I need an area for producing vegetables'? Take this cue and use similar language to ask your questions.

If you are designing your own garden, you can go through a very similar process of asking yourself the right questions. Also, it is important here not to start with too many preconceptions. At first, try to forget as much as possible about what is already in the garden or any decisions you've already made about what should go where. As you brainstorm initial requirements and ideas, allow yourself the freedom to let anything be possible. First get a clear idea of what you want from the garden; the how and where can come later. Sit in the garden on a warm evening with a glass of something good and allow yourself to dream.

We have produced a sample sheet of the questions you might ask a client or yourself. As you get more used to the process this might just be a rough starting point and you can let your client, or your own thoughts guide you as you progress. But at first, if you use these questions, you should end up with a usable, comprehensive brief.

2. The survey

It would be hard to overstate the importance of a thorough and accurate survey of the garden as a basis for your design. Get it wrong and you could find that you have wasted time creating a design that can't be built or, even worse, that you have wasted money on materials that can't be used. However, a basic survey that provides all the details necessary for most projects is easily achievable for the amateur as well as the professional and this section will set out the process.

There is also the option of having a professional survey done. If you are designing your own garden, this is probably not justified unless you have a very large plot. If you are a professional designer you should consider this option, however, especially for large or complex plots with lots of level changes. For your own garden you can always pop out to check a measurement or add a level to your plan but this is not always practical when you are working for a client, so the survey needs to be accurate and complete enough for your needs from the start. At the time of writing a survey of a typical, average sized private garden in the UK might cost between £500 and £1,000.

If the garden is not too large or complicated, it is very feasible to produce your own survey that includes everything you need to produce a design. A good starting point is to obtain a 1:500 scale map of an area that includes the property and these can be found online and purchased as a PDF. If you are not using software, you can have these maps posted to you and then you can either work from them at 1:500 or take measurements and adjust to whatever scale you are using for your drawing.

SCALE

The scales most typically used in garden design are as follows:

1:10 or 1:20 – Construction drawings of features such as walls, pergolas and ponds. 1cm on the drawing represents 10 or 20cm in the garden.

1:50 – Outline designs for very small gardens, planting plans of individual beds or areas, or construction drawings for larger features. 1cm on the drawing represents 50cm in the actual garden.

1:100 – Full outline plans on A3 or A1 paper of typical residential gardens. Planting plans of smaller gardens might also be at this scale. 1cm on the drawing represents 1m in the actual garden.

1:200, 1:500 – Outline plans on A3 or A1 paper of larger gardens. 1cm on the drawing represents 2m or 5m in the actual garden.

If you have obtained a 1:500 map of the property you should use your software to change the scale to something that will fit nicely on a sheet of A3. For example, you could enlarge by 500 per cent to end up with a plan at 1:100. You can then print this out and clip it to a drawing board ready to take on site and put your own measurements on it. It is a very useful start and gives you confidence in things like the angle of the boundaries to the house which are often not nearly as parallel or perpendicular as they might look to the naked eye. However, you will still need to check all measurements as the way mapping services record boundaries may not reflect the actual fences, walls and hedges you will be working with. The following process is the same whether you have obtained a map or are working from scratch.

You will need a good heavy-duty short tape measure, usually 5m or 8m long, a long surveyor's tape measure, usually 30m or 50m long, and some sort of pegs or stakes (we use steel fencing pins/road pins). A folding square can also be useful for checking that you are measuring at right angles to the house or boundary.

Starting with the house

To explain the process, we will work through the survey of a rear garden. To produce a survey of the whole plot you would repeat the process for each side of the house and join the surveys into one overall plan. A good first step is to measure across the back of the house. Buildings are easy to measure. The walls are usually perpendicular or parallel to each other and it is easy to assign fixed points from which other parts of the garden can be measured. Mark the first corner of the house on your plan as a capital A then

Corners of the house – A, B, C etc.
Doors – d1, d2, d3 etc.
Windows – w1, w2, w3 etc.
Steps – st1, st2, st3 etc.
Lean-to shed – sh1, sh2, sh3 etc.
Down pipes – dp1, dp2

A:d1 = 0.62m
A:d2 = 2.85
A:d3 = 3.82
A:d4 = 4.66
A:w1 = 6.52
A:dp1 = 6.85
A:w2 = 7.35
A:w3 = 9.82
A:dp2 = 9.10
A: st2 = 10.26
A:B = 11.27

B:d5 = 0.84
B:C = 1.17

C:D = 4.15

D:sh1 = 1.16

st1:st2 = 4.04
st2:st3 = 1.11
st3:C = 1.02

sh1:sh2 = 1.35
sh2:sh3 = 1.82

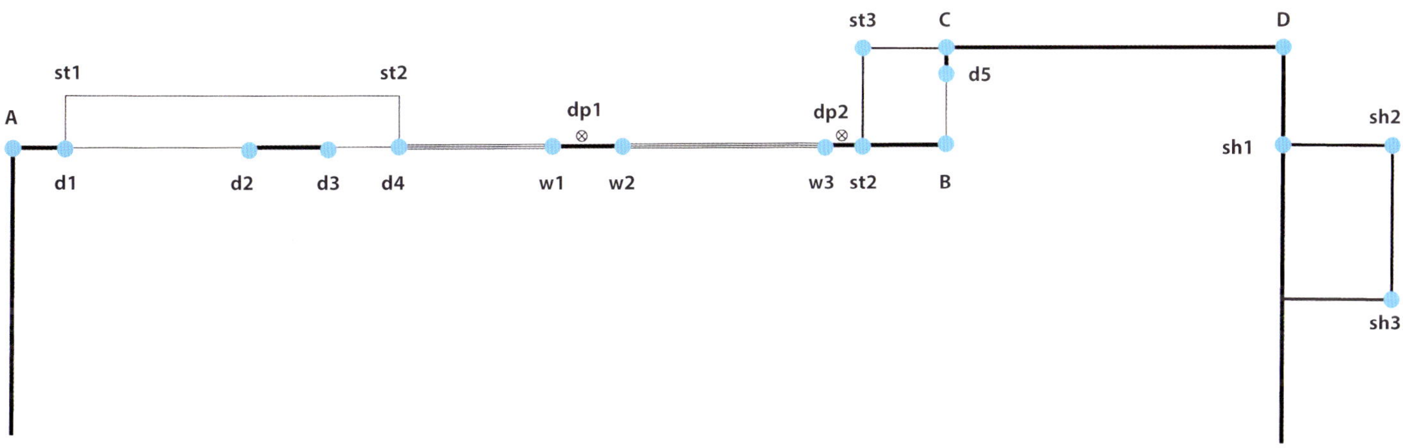

Measuring the house.

measure along the wall marking the start and end of each window (w1:w2, w3:w4, etc.), each door (d1:d2, d3:d4, etc.) and any other important points such as a down pipe, as a distance from A. We measure all points as a distance from A, rather than just measuring from one point to another, so that any error does not make every subsequent measure incorrect. We measure the position of the doors and windows so that when we come to creating a design not only can we avoid obstructing them, but we can also consider important views from the house. It can also be very effective to line up the layout of parts of the garden with the important doors and windows, as we shall see in later parts of this chapter. When you reach the next corner point, mark that as B and continue the process until you reach the far end of the house.

Next we measure from each end of the house out to the boundary. If you are certain that the garden is rectangular, perhaps because you have obtained a map of the property, it should be easy enough just to measure to each corner of the garden, labelling each point consecutively as a roman numeral (I, II, III etc). If the garden is not rectangular you will need to measure some points by triangulation.

TRIANGULATION

Triangulation is a useful method to use to measure points in the garden or on the boundary which are not perpendicular to another boundary or the house. It can also be used to measure a number of points on a curved boundary. Two known points, which you have already measured, are needed, from which you can measure the distance to the new point. Two points on the house are usually best for this and it is better if they are fairly far apart, for instance the two end points of the back of the house. We then measure from each of these points to the point in question, for instance a fence post. The measurements can then be drawn on the survey as two circles whose radii are the measurements you made. The point where the circles meet will be the position of, for example, the fence post. The circles you draw on the plan will in fact meet in two places, but usually it is obvious which of these points is the correct one.

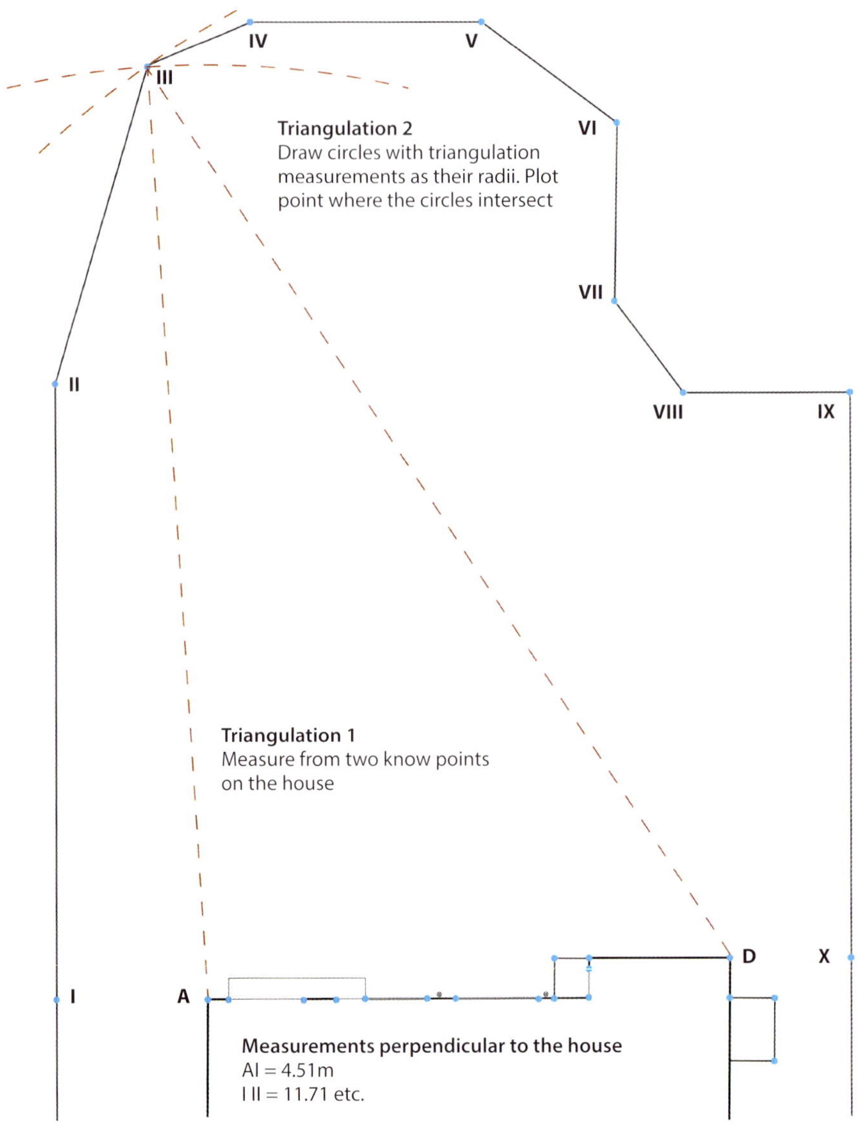

Measuring the boundaries and triangulation.

Adding significant features

Having established the house and boundary, we then measure the position of important features such as outbuildings, patios and ponds. We measure everything that may stay in the design as part of the new garden. But if, for example, we know that we will definitely be replacing and redesigning the patio, there is probably no need to include it in the survey. If the feature, for example a shed, is perpendicular to a boundary we can position it by measuring along the boundary, then out to a corner of the shed. We then measure the dimensions of the shed. If the feature is not perpendicular to the boundary, we will have to find the position of at least two corners of the feature. Triangulation will probably be the easiest way to do this.

The position of important trees should also be located and the easiest way to do this is by triangulation. The width of the trunk should be measured as well as the canopy. Ideally you would measure the canopy in two or more directions as they are not usually very circular. It is sufficient for our purposes just to measure the canopy by placing a pin or getting somebody to hold the measure at one edge of the canopy, judging by eye, and then measuring across the centre to the other edge, again judging by eye where that point is.

More irregular features such as a pond or rockery are trickier to measure but can usually be surveyed accurately enough for our purpose using offsets. As a relatively straightforward example we will imagine that we want to measure a bed that we have decided will probably remain in the final design. By taking a series of points along a boundary, regularly spaced, and measuring from them to the edge of the bed we will end up with a reasonable approximation of the shape of the bed. If the bed is along a fence, we could use the fence panels as conveniently spaced points to measure from. Otherwise we will have to mark our own points, choosing a spacing to reflect the complexity of the shape, and marking them with fencing pins or pegs. We then measure the perpendicular distance from the boundary to the edge of the bed at each point. We can probably get away with judging the right angle by eye but if in any doubt use a folding square.

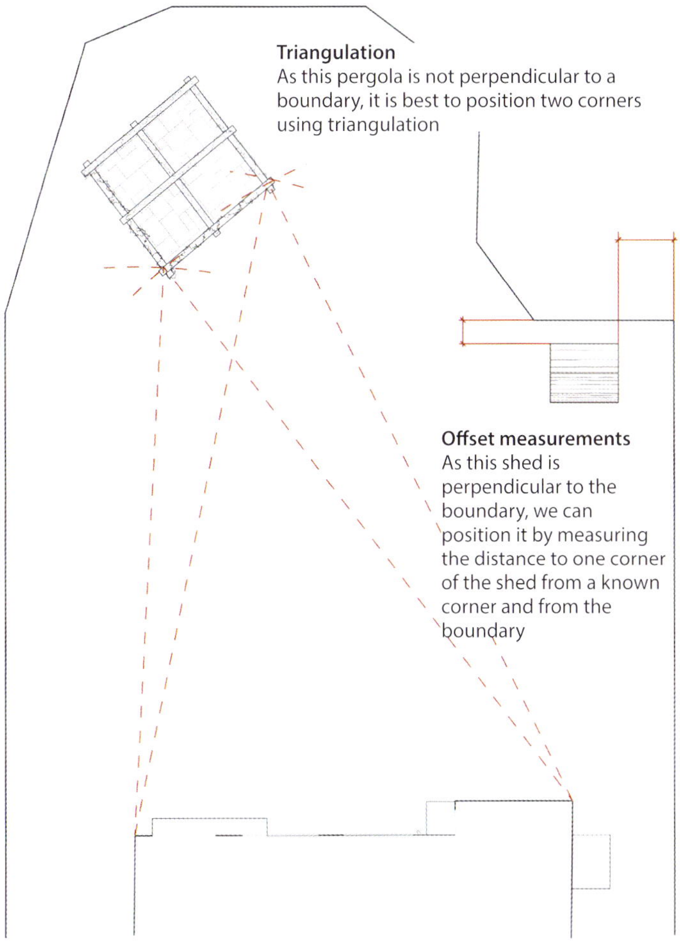

Measuring the position of a shed using offsets and a pergola using triangulation.

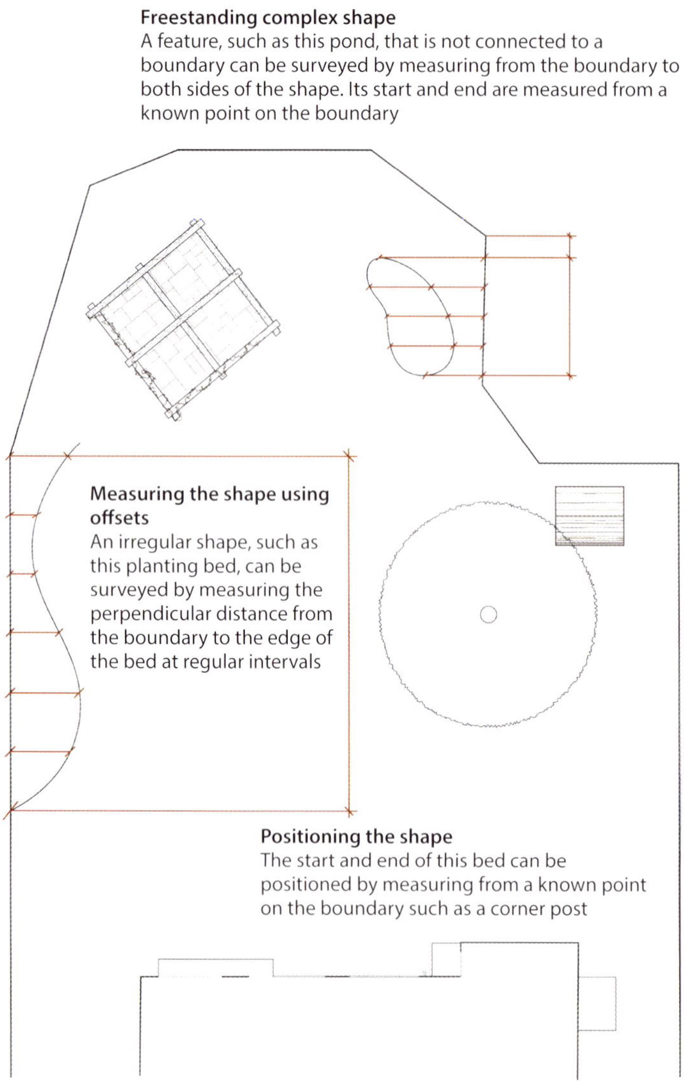

Surveying irregular features using offsets.

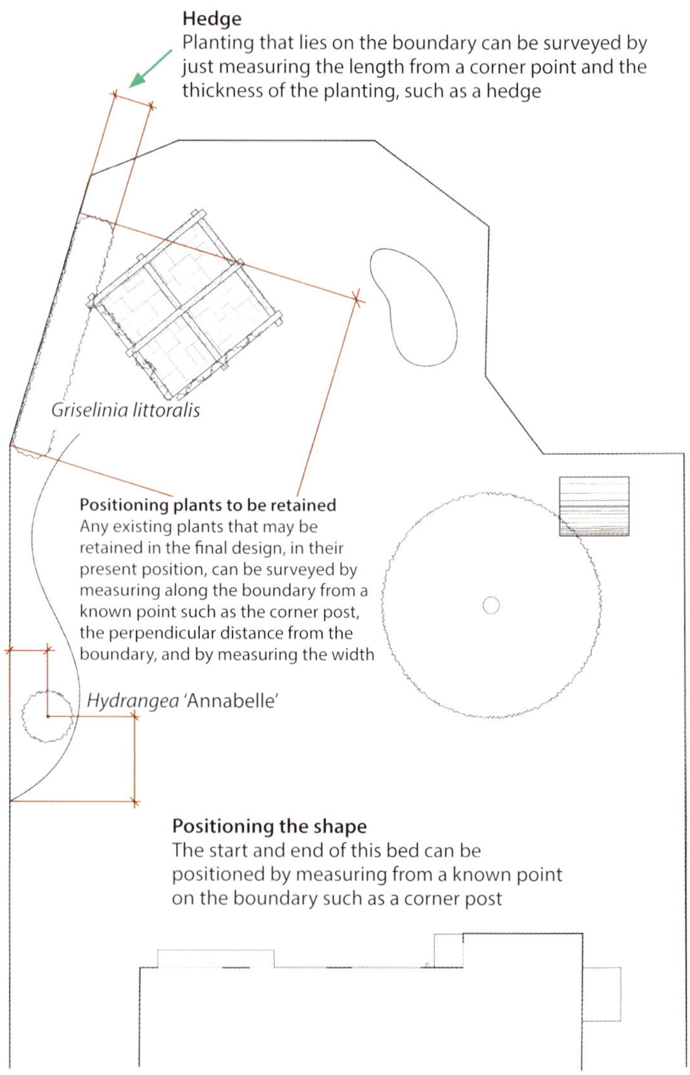

Adding other planting.

At this point it might be a good idea to draw the survey information collected so far onto a scale plan which you can print out or photocopy to use to record further planting and levels information. If you are not working on your own garden, this might not be feasible. Instead, you could quickly copy the outline of the property and boundary onto a new sheet of paper so that your work so far does not become too cluttered. If the survey is quite simple, you could just use different coloured pens for the planting and levels information on the same sheet.

Adding planting

We now consider what other planting, apart from the trees already recorded, we might be keeping and using in the new design. For plants that will definitely be kept but will be moved, we just need to list and photograph them. If a plant might be kept in its current position, it needs to go on the plan. Usually, the easiest way to record this position is by measuring along a boundary from a known point and then measuring out to the plant, i.e., as an offset. Otherwise we may have to measure its position using triangulation. If all of the planting in a bed is being kept as it is, we will probably only need to make a list of what is in it and to take a few photographs.

Changes of level

Lastly, we need to measure changes of level across the garden. This can be a bit trickier than the measurements we have covered so far, and it can be tempting to skip this stage, particularly if the garden appears to be fairly level. However, it is extremely rare that a garden plot is so level that the design doesn't need to take at least some level changes into account. There are important consequences for things such as the need for a retaining edge for a lawn or bed, the flow of rainwater across a patio or how loose gravel is to be retained on a path. Many of these problems could be dealt with as they occur during the build phase, but that could result in unforeseen cost or even major changes to the design. Far better to have accounted for these issues beforehand and, if you are working for a client rather than yourself, far more professional.

For a small, uncomplicated plot, it may be sufficient to use the following method using a spirit level, 1m plank of wood and a tape measure.

1. Starting from a known point at the top of a slope, place the 1m plank facing down the slope parallel or perpendicular to something known like a boundary or wall of the house.
2. Make the plank level using the spirit level and measure the distance from the top of the plank to the ground.
3. Mark the point, move the plank down to that point and repeat until you have reached the bottom of the slope.

All these points and changes of level can then be recorded on your survey sheet. Other important level changes can be measured on a point-by-point basis such as a step up from patio to lawn, a set of steps, the height of a wall or the step down out of the house. Use different coloured pens so as not to get slope measurements mixed up with other spot measurements. This method is sufficient for a lot of gardens, particularly if you are not working for a client. You might even get away with using the change in height of a series of fence panels along a boundary to get some idea of the slope of the garden. But you should start with at least some idea of the slope of the garden before you start your design.

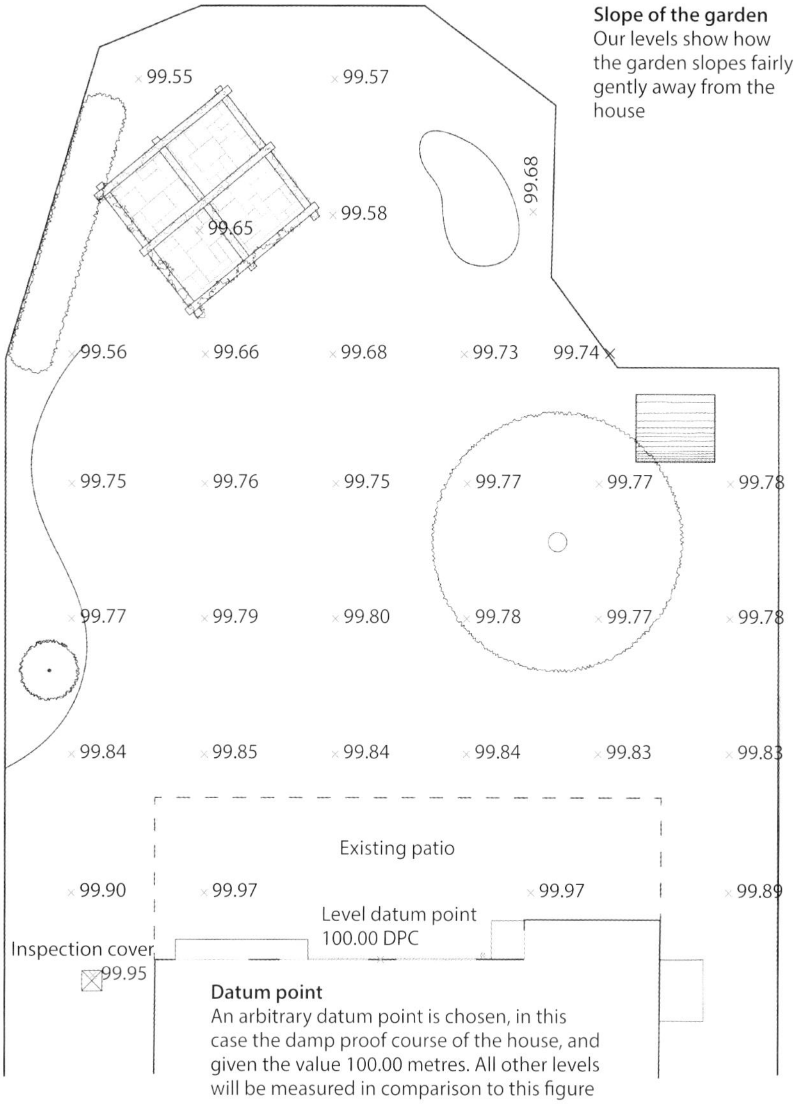

Levels

For a more complex site, or if you are producing a professional design, you will need a more detailed and accurate method. A very complex site will require you to employ a surveyor. Otherwise a lot can be achieved with some basic surveying equipment, either bought or rented. The traditional dumpy level and levelling staff are still widely used and not very expensive. You will probably need some training though or at least some good instructions and some practice before you start using them professionally. It is also very much a two-person job. A rotary laser level can be used instead of the optical dumpy level and is a little less fiddly but is a similar process that still needs some practice or training. Probably best of all if you are a professional designer is to invest in a newer type of level that uses a long fluid-filled tube to measure level changes. Taking the levels using one of these can be done at the press of a button, can easily be used by a single person, and for the size of plots that garden designers typically deal with they are ideal. They are a bit more expensive than other options, but you may be able to rent instead. Brand names on the market at the time of writing include NivComp or ZipLevel.

All the above methods will require you to establish a datum, a point from which all the other levels are measured from. This needs to be something that will not be altered or removed during the building process. The sill of a door or the damp proof course are often good datum points. Points across the rest of the garden are then measured and the change of level from the datum point recorded as a + or – number of, usually, millimetres above or below the datum. These points usually form a regularly spaced grid across the garden. Other spot measurements such as the height of a wall, or a series of steps are also recorded.

If you are using CAD software, it is a good idea to place the main drawing of the house and boundaries on one layer, the trees and plants on another, and the level data on a third. Then you can turn the visibility of any layer you are not using off, so as to keep things uncluttered. A similar effect can be achieved if you are drawing by hand using sheets of tracing paper. The hard, routine work is now done, and you can move on to the more interesting and creative stages.

3. Site analysis

The next step involves a careful analysis of both the potential of the site and its limitations. Ideally this will be done on a different day from the survey, giving you the chance to draw and print a scale plan to write your observations on. Otherwise, if you are working for a client and it is not convenient to return you could just work on another print of your basic map or quick sketch of the boundaries and buildings.

Views

Start by considering the key views from the house. Position yourself in front of some of the key points from which the garden will be viewed, perhaps the patio doors from a living room, a large kitchen window, or think about how it will look from a first-floor balcony. If the property is right on the coast, then it is likely that the view of the sea is going to be the first priority. The first impulse might be to open this view up as much as possible but that is not always the most effective approach. Carefully placed trees or large shrubs might frame the view to much better effect than just leaving it all open. Mark on your drawing where you think the best views of the sea are. What about other features that might help create the best sea view? A pier or island might give a sense of perspective or point of interest to the view. Perhaps the sea is best viewed not from the house or main terrace but from a point further down the garden. You might then start to think about how a series of views might be created so that you catch glimpses of the sea as you move down the garden until it finally opens out into the best, most open view. This sort of 'narrative' approach is probably best suited to large gardens but if the potential is there it can be a very atmospheric effect.

Perhaps, particularly if the garden is set further inland, the sea is not visible. What other views might be worth emphasising? A distant line of trees might be framed by new trees in the garden creating a sense of space and of the garden being linked to the wider landscape. The garden might border a picturesque meadow that can be visually 'borrowed' as part of the garden by creating a break in the fence line with a false, non-opening, five bar gate. Mark these views and initial ideas on your site analysis drawing.

Of course, there may also be views that are not desirable at all. An overlooking next-door window or ugly building might need to be screened or distracted from as much as possible. It may be that there are no good views at all out of the garden and your design might need to create an enclosed space with all the attention drawn inwards into the garden itself. Mark all these problematic views on your analysis. Thought must also be given, however, to the effect any screening such as large trees, hedges or fences will have on neighbouring properties. There are planning regulations to consider that govern things such as the height of fences but beyond this it is important to show good consideration and try not to block the sun or good views from neighbouring gardens. Other statutory constraints should also be researched, such as Tree Preservation Orders (TPO) or Conservation Areas.

Practical considerations

Everyday practicalities should also be explored on site and your observations recorded. How will people typically move around the garden? What are the key routes, for example from the kitchen door to the bins or from the driveway to the front door? You can quite often see in parks, worn routes across the grass where people are not using the paths but heading more directly to where they want to go. These 'desire lines' often show where the park's designer has not anticipated how the park will be used.

Whilst you are thinking about how the garden will be used, you should also consider what practical features are needed and where they might best be situated. Many of these should come from your brief but check that nothing has been missed. Most gardens need a spot to hide things away in storage, a place for bins, perhaps a log store or compost heap. Think where these should best be placed. A raised herb bed at the far end of the garden is a lot less likely to be used than one right outside the kitchen door.

Sunshine and shade

A herb bed is also going to need to be placed in full sun, which brings us to our next consideration, shade in the garden. North should be marked clearly on your site analysis. It is the first source of our ideas of how light and shade will work in the garden. A south-facing garden will have a different relation to the sun than a north-facing one. An east-facing border will catch the morning sun but may be shady in the afternoon. This is going to be important as you decide on the correct type of planting or the placing of sitting areas. Mark any significant areas that are in the shade of hedges, trees or buildings.

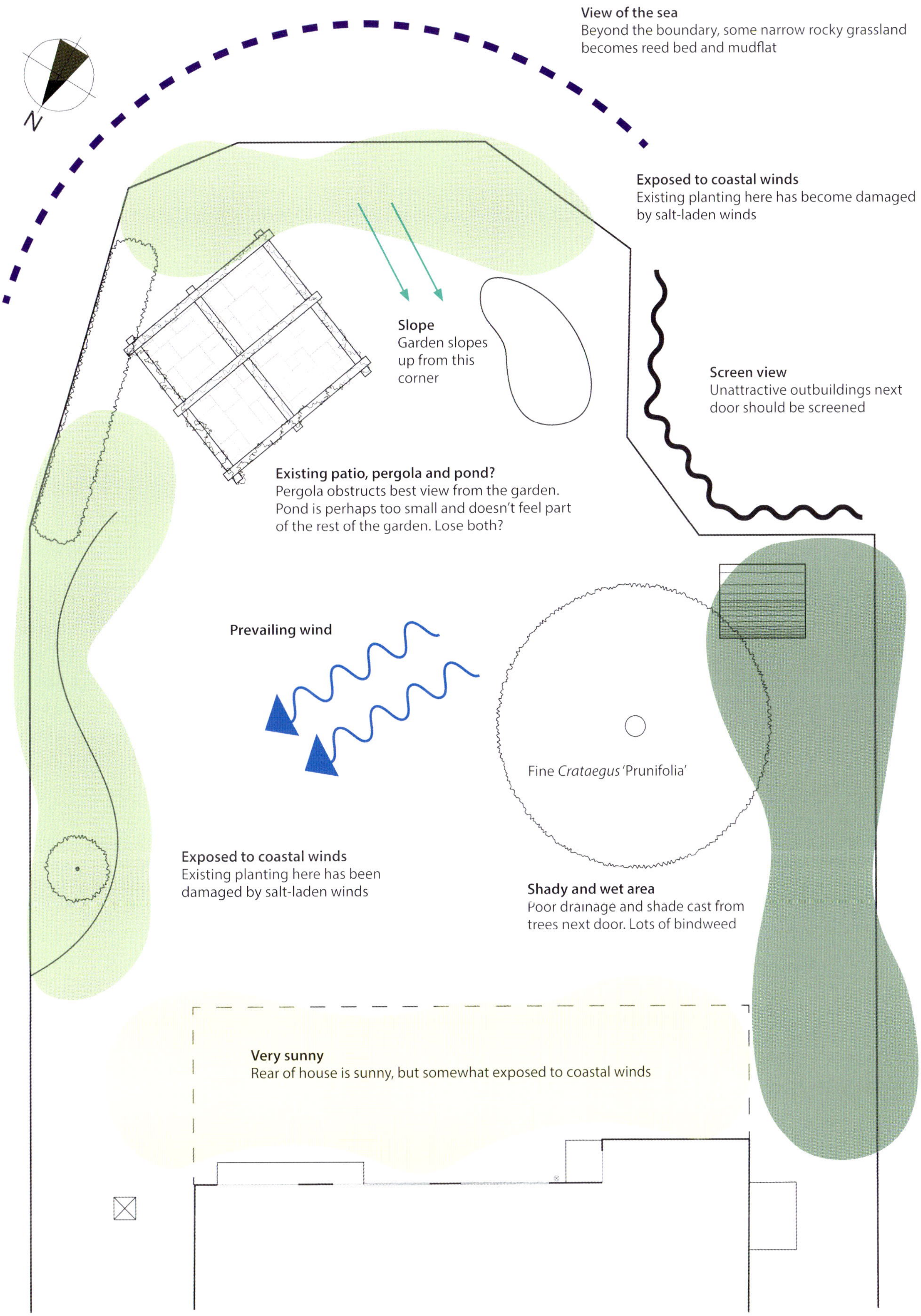

Site analysis.

Shelter

The quality of the light is often one of the key features of a coastal site and you are likely to want to place a sitting area in a good sunny spot. But it will also need to be somewhat sheltered, which brings us to the next consideration, one which is very key to coastal gardens, the wind. A fresh, bracing sea breeze might be one of the great pleasures of being on the coast but it brings serious challenges to creating a successful garden. Places on the coast tend to be windier than low-lying inland areas because the sea provides less friction to slow down the incoming wind. Also, these winds coming off the sea carry salt with them and plants that are naturally found along the coast have adaptations to deal with these conditions. We will see in later chapters how to choose and use plants that have adapted to coastal environments, and we will consider how best to work with the layout of the site to create shelter and to reduce the effect of these winds. So, an important part of the site analysis is to establish how the garden interacts with the wind.

Firstly, you should get an idea of the prevailing wind direction. For the majority of places in the UK this will be from roughly the south-west. But there are regional topographic variations that may mean the average wind comes from a different direction. Also, when you are considering a single garden rather than a wider location, its surroundings may make a great deal of difference. The wind may bounce off the side of a neighbouring large building or be filtered by a copse of trees. The average wind direction may also vary according to the season. If you are planning your own garden and you have only recently moved in it may well be worth living with it for a year before starting your design, so as to have a better idea of local conditions. If you are working for a client, you should be sure to ask them what they have observed about windy spots in the garden. Existing planting should also give you some clues about what is going on. You will often see that plants reach a certain height and then all foliage above that height is scorched or has died off completely. Or you may observe that all the planting between two buildings is showing evidence of severe wind and salt damage despite seeming to be sheltered from the prevailing wind direction. In which case, it is likely that salt-laden wind is being funnelled between the buildings, creating a wind tunnel. Hard barriers such as fences or walls may cause eddies in the wind, and you should look for evidence in the existing planting as to where the wind deflected by the barrier is swirling back into the garden. Note all your observations about the wind onto your analysis sheet.

Soil

Next you should investigate the soil around the garden. Has the plot been a garden for a long time or is it part of a new building development on what was previously waste ground? Gardens of newly built houses are quite likely to have only a thin layer of topsoil, imported by the builders to cover soil that is likely to be of poor quality, or worse, to have been used to dispose of materials from the build. Find some spots around the garden where you can dig exploratory holes to give you an idea of what you will be working with. Soils along the coast are often quite light and sandy but can be quite fertile in places, explore and make notes on your analysis sheet. You will also need to get an idea of the pH of the soil. Ideally you would use a kit to measure this, taking samples from a few spots around the garden as it can vary. These kits can be bought at garden centres. Taking note of what plants are thriving in the garden and in surrounding gardens will also give you a clue, for instance heathers (*Calluna vulgaris*), camellias or rhododendrons would suggest that the garden has largely acidic soils. It is a good idea to look around at neighbouring gardens to give you some ideas of plants that do well in the area.

Typical scorch damage caused to leaf edges by salt-laden coastal winds.

Large planting is being used here to filter the effect of the wind tunnel created between the house and boundary wall at this sea front property.

Final thoughts and photographs

Finally, you should walk around the garden again and make some more subjective notes. Is there a part of the garden that just feels cosy and makes you want to linger? Perhaps there should be at least a bench in that spot? Do other areas just feel like somewhere you want to move quickly through in order to reach another spot? Perhaps you might develop this area to make it more interesting or perhaps you may want to work with that feeling and make the most of a distant focal point, making it a more intriguing and obvious destination. What about other senses? Do you find yourself in a spot surrounded by bird song or the murmur of bees? Perhaps you might develop this part of the garden as a haven for wildlife and the feeling of nature. Is there an intoxicating smell of honeysuckle in a hidden corner of the garden? A good spot to site an arbour seat for an evening gin and tonic, perhaps. Record these observations and be as vague or poetic as you like. You don't need to share these impressions with anybody, but they may be the seed of a creative idea for the design later on.

Finally, if you are working for a client and not on your own garden, you need to take plenty of photographs. In these days of digital cameras or even just your phone there is no need to be stingy with the number of photos you take. You can always delete them later, but we find that however many we take, we almost always end up needing a photograph of something that we forgot to take. Take long shots of the garden as well as close-ups of existing planting or features; take photographs of the key views; take photographs of the house and of the existing planting – anything that could be useful later to help you produce your design.

4. The outline plan

Now you should have a base plan of the house, boundaries, key features and levels drawn up and printed out to scale, likely 1:100 or 1:200 on a sheet of A3. You also should have your site analysis and your brief ready and to hand. You are now ready to start the most exciting part, the actual design. What is now a plot of land with its own problems and potentials, reduced to notes and measurements on a piece of paper, is going to be transformed, through your imagination, into a solid plan from which, with some good physical labour, will be turned into a reality. The initial brief of requirements and desires, either your own or that of your clients, will combine with the realities of the site and moments of inspiration to produce, finally, a real, living, growing garden that might be nothing like your initial thoughts but will, nevertheless, be a thing of delight for many years or even decades to come.

If you are designing your own, dream coastal garden, this is likely to be one of the great, creative projects of your life. If you are working as a garden designer, you probably find that the pay is poor and there are all the stresses and irregular hours that come with being self-employed, but this creative act of turning ideas and desires into an actual garden is the heart of what you do and that is not a bad way to earn a living at all.

Throughout this process you should make sure that you are constantly referring to the initial brief, particularly if you are working for a client. It is easy to get so caught up in an exciting, creative idea for a cascade of coastal meadow planting that you forget that the children want a football pitch. A good starting point in the brief is the general desired style. Is a more formal style desired, with straight lines and symmetry? Or something more relaxed and natural with flowing curves?

Think about how this style will work with the house. A well-designed modern house that uses concrete, glass, and strong, simple lines is going to suit a simpler layout, perhaps with wide planting beds, straight lines and landscaping material that reference those used in the building or in the surrounding natural landscape. Done well, this architectural style has a sensitivity to its surroundings that can work fantastically well in a coastal setting, and you should complement this with your design for the garden, using as many local materials and plants that grow in the area as possible.

An older property may suggest a more relaxed approach and many people want their gardens to be a natural, flowing space, free from the straight lines of the built environment. This is a very valid desire, but it is worth bearing in mind that it is hard to make complicated, curvy shapes work in small gardens. In these gardens you might use a circle, or part of a circle, to draw attention away from the boundaries or use straight lines but soften them with your planting.

It is usually a good idea in a smaller garden to draw attention away from the boundaries to help give a feeling of more space. One exception could be a courtyard garden where the walls and climbers on the walls are part of creating a cosy, enclosed atmosphere. One of the most common mistakes in gardens is to just put a narrow border of planting around the edge of the garden, creating a boring effect that makes the garden feel smaller than it has to. A circular lawn is a good trick to get away from this, or just a bit of planting between the patio and the lawn, framing a particular view. But, as with everything in design, this is not a hard and fast rule. It may be that in a small garden you need to use every bit of the space in the middle of the garden for seating and entertaining, or for a children's play area. It is very possible with clever use of containers or other focal points to create a pleasing effect that leaves the centre of the garden clear.

Contemporary architecture calls for a more modern style of planting such as drifts of perennials and interesting textures such as ornamental grasses.

A curvilinear design works in this large garden but it could be too fussy in a smaller space.

Although there are conventional wisdoms and tricks and tips that produce a rough consensus among garden designers of how to approach different types of gardens, none of this should be considered set in stone. What may seem common sense to a generation of garden designers may become an uninspiring cliché until an imaginative designer comes along and breaks the rules a bit. Much more important is the process of thinking about the challenges and possibilities of the particular garden you are working on. So, feel free to ignore any of the general advice about solving particular problems in garden layout, as long as you have recognised and thought about how you are going to deal with the issue. This applies less to plant selection where the conditions, particularly along the coast, do limit your freedom of choice quite a bit. But even with the planting there is room for experimentation and innovation. Thirty years ago, not many gardeners were thinking about using wildflower seed mixes in coastal gardens.

You should now start to sketch some ideas for the general shapes and layout of the garden. You want to feel as free as possible at this stage to experiment. Nothing is set in stone. You can make mistakes and try lots of different approaches until you find one that works. Print out a few copies of the base plan to work on. Alternatively, you could use a roll of tracing paper, working over

FUNCTIONAL SIZES

You should make sure that things such as the patio or paths are drawn at a suitable size right from the early stages of your design. You don't want to find out only at the end of the process that what you have drawn will not fit the furniture you have.

The patio should be in scale with the house. Basing the size of the patio on the size of the living room is often a good start. Then consider the patio furniture you have or plan to use. How many people should be accommodated? If you can, lay it out on site using builder's site marker spray paint, or pegs and rope. Allow enough space for chairs to be pushed back from the table.

Small patio – a good size for a small, cosy patio for a bistro table and a couple of chairs should be from about 1.8 × 2.4m to about 2.4 × 3.7m.

Medium patio – a table for four people should be on a patio measuring at least 3 × 3m. Six to eight people could be accommodated within a 4 × 4m space.

Remember to allow space for sun loungers or for a barbecue if required.

Paths – for main paths, a width of about 1.2m works well. It allows two people to pass comfortably and makes a good, visual impact on the design. Less used paths could measure about 90cm across and the minimum practical width, perhaps in a more functional area, would be about 45cm.

the base plan, sketching one idea then moving along the roll to sketch the next one. If you are working for a client, this is a great way to show them your thought process and, in our experience, clients really like being shown this. Start by noting down anything major from the brief or analysis that has to be included in the design in a definite position that is unlikely to change. For instance, it might be that there definitely must be a patio, adjoining the back of the house that is large enough to accommodate a table for six. You don't want to note the whole brief on the sketch at this stage – that would be too limiting to your initial creativity – just a few of the most important requirements. Then sketch in the shape and size of a couple of the largest features in the garden. This could be the patio and the lawn, although as we are thinking about coastal gardens, we may decide to create an area of mixed hard surfaces, gravel and gravel planting instead of lawn as they can be hard to establish well in very exposed sites. There are themes you might use as a basis for the layout, and we have illustrated some examples of these in this chapter. A design is usually stronger if it has an overall theme. There could be more than one theme, particularly in a larger garden, but the areas should probably be linked in some way, for example with a repeated planting or landscaping material. You also should feel free to just use these themes as a starting point or ignore them completely and let your pencil roam free and take you where it will. The great challenge and pleasure of design is that 'rules' about what looks good are only guides and can be adapted or ignored completely if you are inspired to do so.

It is important not to forget in this process that even though you are working on a two-dimensional plan, you are designing a three-dimensional space. Keep referring back to the levels data that you recorded in your survey. Walk around the garden as much as possible, imagining how your ideas will look and how they will work practically. The plan view we typically work from is a slightly strange one as it represents a top-down view of the garden. That isn't the way we will normally be looking at a garden. There are some useful ways to get away from too much two-dimensional thinking. The landscape designer Humphry Repton (1752–1818) in his 'Red Books' worked on his designs with watercolour sketches of views of the landscapes that incorporated pasted flaps that could be lifted to show a 'before and after' effect. Even without his painting skills we can do something similar by working with tracing paper over photographs, sketching trees and other features to demonstrate and experiment with views and shapes in the garden. CAD software can also be very helpful. The 3D modelling software SketchUp provides a particularly quick and intuitive way to experiment with three-dimensional changes in level and how things like walls and trees might work in the design. But most of the CAD packages used by garden designers will offer some three-dimensional capability. Some are easier to use than others though, and you do need to be careful that struggling with the

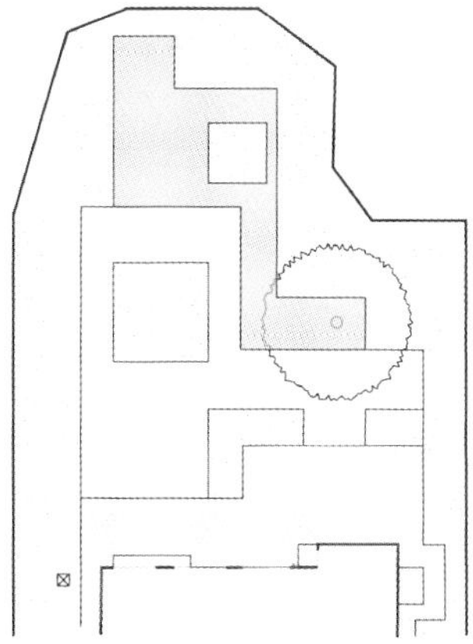

Rectilinear
Straight lines at rectangles create a more formal effect. It can be a good idea to line features up with points on the property.

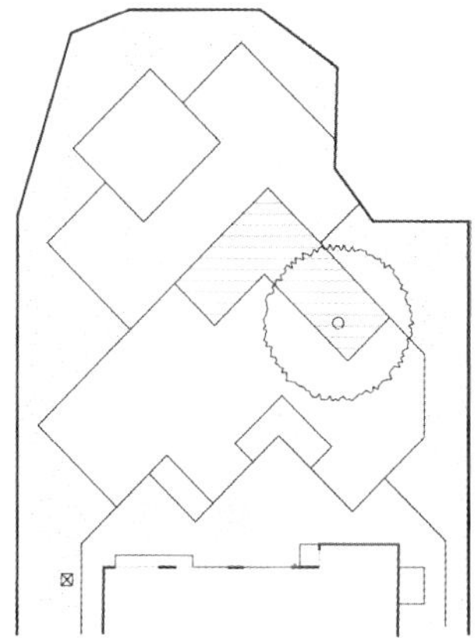

Angular
Arranging the garden using lines at an angle to the house can create a more spacious feeling and direct the attention towards a key view.

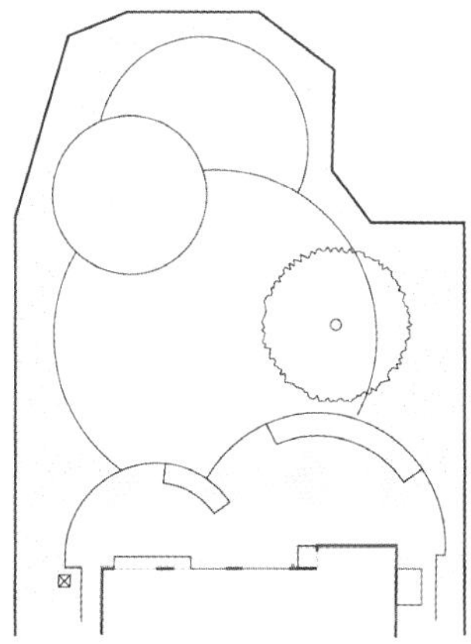

Circular
A series of interlocking circles is a slightly less formal but still well-ordered effect. It tends to concentrate the attention away from the boundaries.

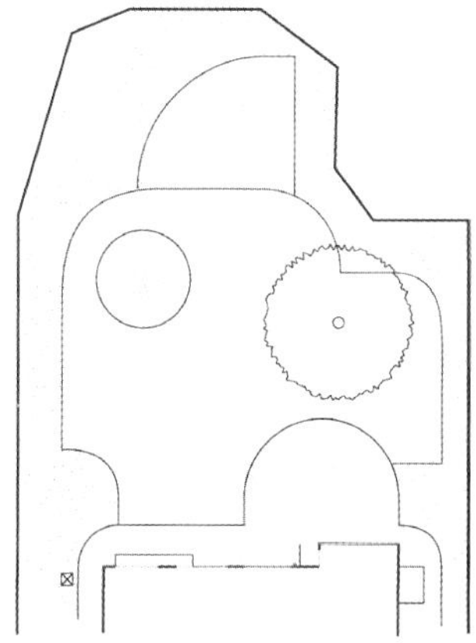

Arc and Tangent
A combination of perpendicular lines, curves and circles allows for a lot of flexibility while retaining some order and formality.

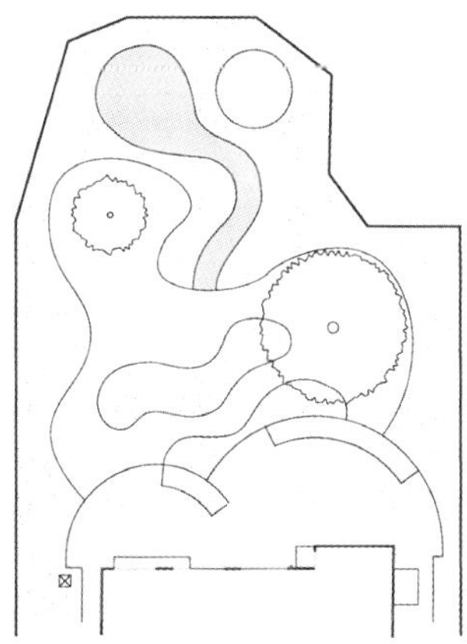

Curvilinear
A design based on more free-flowing, serpentine curves has a much more natural, informal effect.

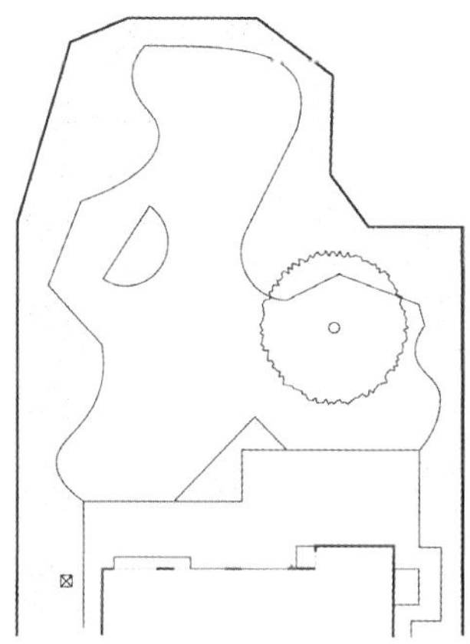

Lack of Design Theme
Often the result of a piecemeal garden development over a time, a garden layout lacking some sensitivity to theme is usually unsettling and unharmonious.

A consistent theme for the way the garden is laid out can create a more harmonious design.

software doesn't interfere with your creative workflow. As long as you keep the levels in mind and use your imagination well, it is perfectly possible to work on just the two-dimensional plan.

The purpose of the outline design process is to generate ideas, and work through them to check that they work practically and aesthetically. The purpose of the finished outline plan drawing is to communicate your ideas for the garden to anyone who is going to help build or plant the garden, and to act as a reference during the construction process. If you are working professionally, it is also your most important tool for communicating your design proposal to your clients. So, it is important that the drawing is clear, and communicates your concept clearly without being cluttered with too much information. For instance, you don't need to include all your levels data. Just an indication of the height of a set of steps or retaining wall will be enough to explain what you are proposing. A suggestion for the general type of planting and colour scheme for a bed is all that is needed on this drawing, perhaps with a few example plant choices. If you are working for a client, then it is worth including some specific hard landscaping material choices so that they can go and look at samples from their local supplier. If it is your own garden, then something more general can just act as a shopping list for now so you can go out and find exactly what you want at a later point. We will be producing much more specific construction and planting details on the later drawings.

If you are working professionally, this drawing is also partially a sales tool and should demonstrate that you have listened to your client's brief and have responded accordingly. It should show evidence that you have used your imagination, expertise and creativity to create an exciting garden that meets their requirements and desires. How closely you stick to the initial brief versus how many of your own creative ideas you use is a tricky balance and one which gets easier with confidence, experience and a good understanding of your clients. Sometimes the brief is very broad, inviting you to bring lots of new ideas. If you have a more restrictive brief but feel strongly that a different approach is warranted, you will need to show your reasoning and here is where your roll of tracing paper documenting your thought process can be very useful. You do need to be prepared for the client to push back against some of your ideas. It is rare that you will present a design to a client and it be accepted completely as is. You must be prepared to go away and alter your design in reaction to comments and continue to do so until your client is completely happy. How you charge for this is up to you. We have always included a certain amount of alteration in the initial quotation for design services. Other designers charge for each redraw.

Once you have a final outline design which everyone is happy with, you can proceed to the next stage. We now need to produce drawings that show how built elements of the garden such as patios, steps, pergolas and raised beds will be built, and how the garden will be laid out, with clear and exact measurements and levels. We will also need a plan to show what we are going to plant and where. If you love gardens you are probably going to find the planting plan the most exciting part of the design process, but getting the details of the hard landscaping work to make the garden both useable and attractive is a very satisfying process as well.

5. The setting out plan and construction details

Now that you have an overall outline design for the garden that you are happy with, it is time to start thinking about how it is going to be constructed. The setting out plan is a version of your outline plan with measurements to help you, or the landscapers you employ, to position and set out all the non-planting features on your design. The construction detail drawings will show how the built elements such as patios, pergolas or steps are going to be constructed.

If you are going to be doing all the work yourself, you may not need all these drawings. If it is a very simple design, you may not need any of them. But if you are going to employ landscapers these drawings are going to be a very useful tool to communicate your ideas to them and to ensure that any quotations you get are as accurate as possible. If you are getting a number of competing quotations for the work, these drawings, along with your specification document, will ensure that all quotations are for the same

Designer Paul Williams makes very clever and interesting use of coastal found objects on a bamboo frame.

A worked example

DEVELOPING THE OUTLINE PLAN

Here we will be working through an example, illustrating the process to arrive at an outline design for an imaginary coastal garden. It will be a rear garden, of a reasonable size to give us scope to demonstrate lots of different ideas, running down to a harbour's edge where the coastal setting is going to present a considerable challenge as well as some great views and creative inspiration. What follows is a step-by-step illustration of how we arrive at a finished outline design plan for that garden.

In reality, you will probably find that you experiment with more basic themes and layouts than in this step-by-step example before you find something that works. Sometimes an idea seems promising but doesn't work when you start drawing the features to scale and taking changes in level into account. If you get stuck for ideas at any stage, an internet search for gardens of the size or style you are working on can help. Good creative work often involves taking inspiration from lots of work that has gone before. By the time you have mixed these ideas taken from other gardens, applied them to your particular plot, and filtered them through your own creative instinct you can still end up with something that is interesting, original and a true expression of your own interpretation of the site.

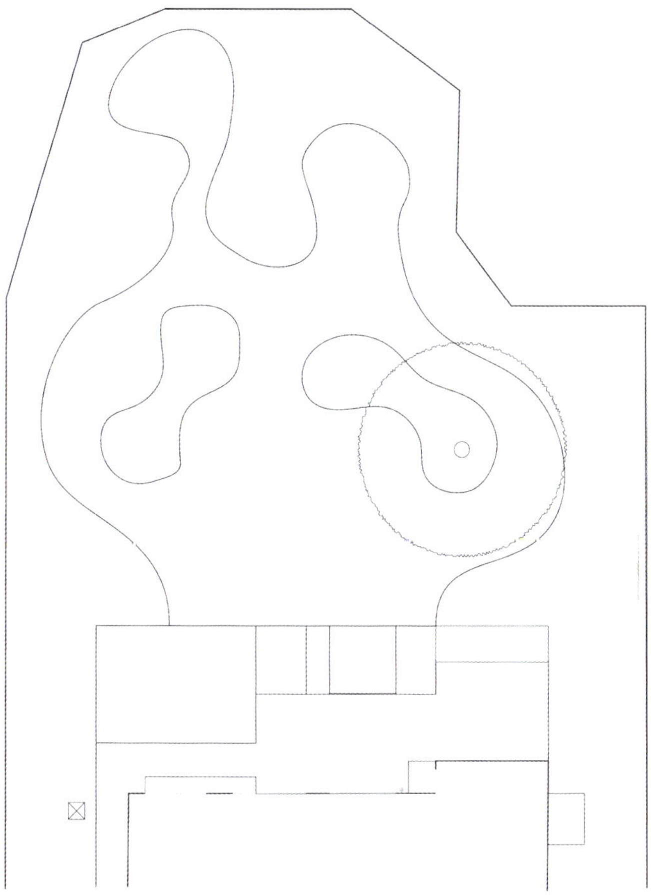

Step 1: Design theme

We decide on a combination of two themes: more formal and rectilinear close to the house; more natural and curvilinear away from it. An existing tree is left on the plan as it will definitely stay.

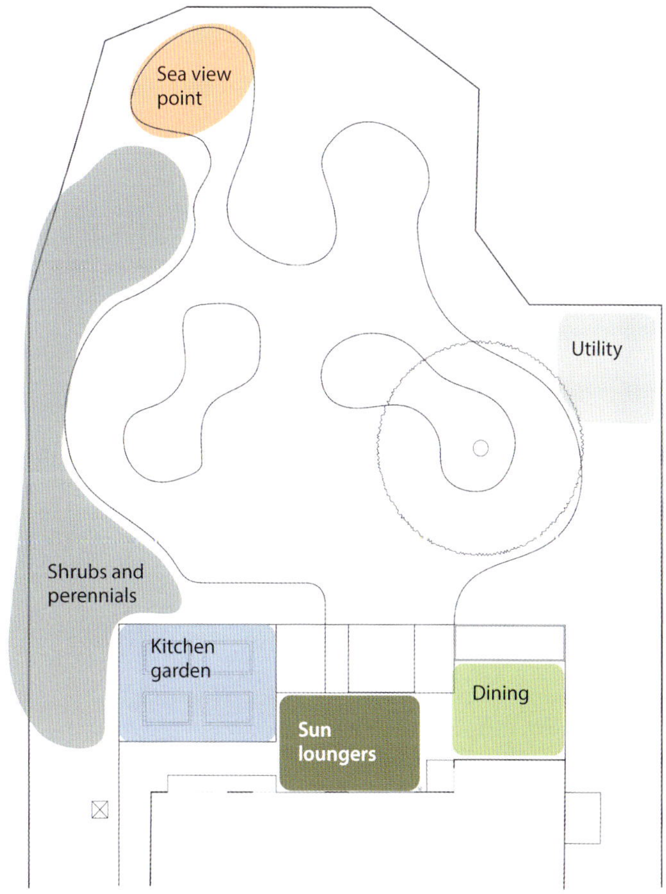

Step 2: Features

We add desired features from the brief, including a well-sheltered productive area, a utility area, and sundowner/viewpoint sitting area.

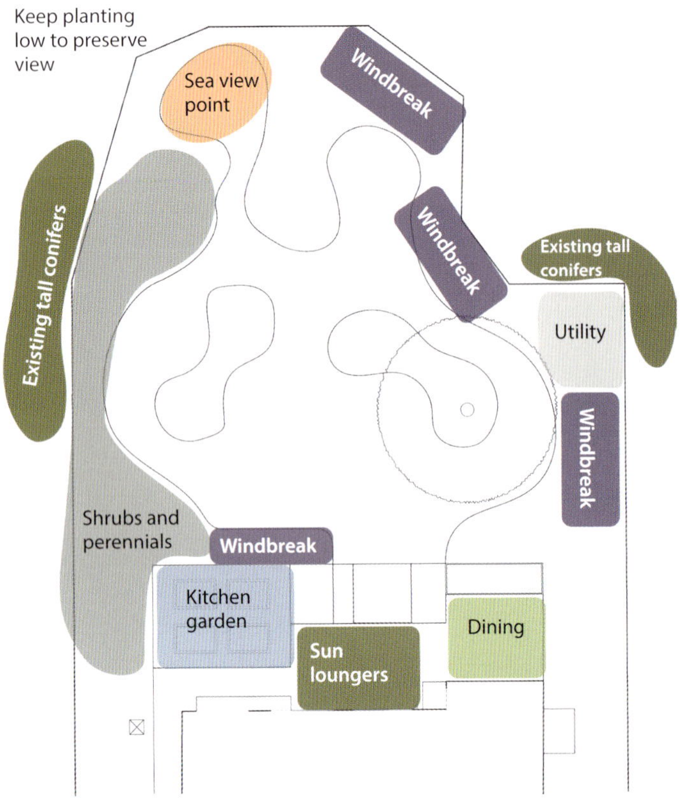

Step 3: Shelter and views

We are thinking about dealing with the prevailing winds and framing key views. Also noted are some tall conifers in neighbouring properties that will protect the garden from the salt winds but also block views.

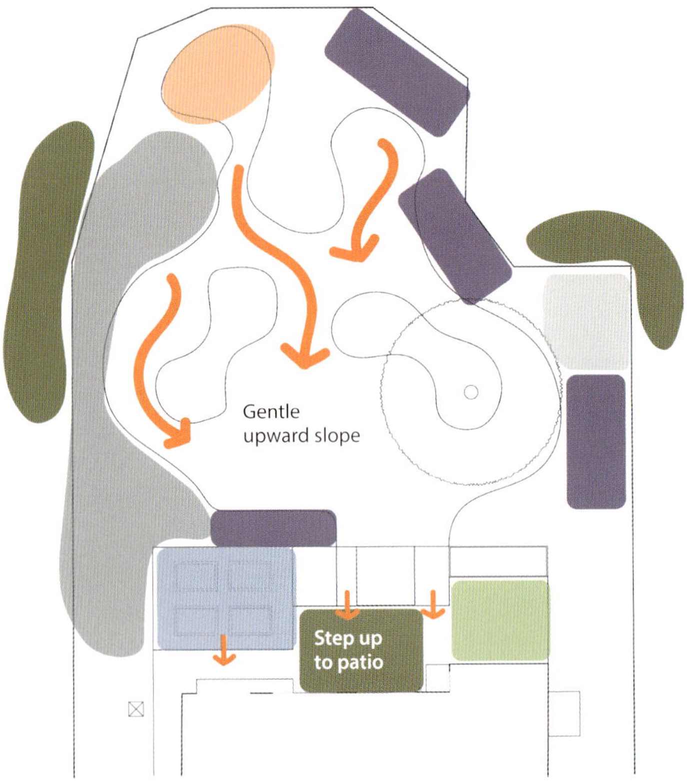

Step 4: Levels

Reviewing the levels we measured during our survey, we can see that this layout should be able to be graded as a gentle slope up from the far corner and a small step up on to the patio. We will not need to create terraces or other structural changes of level.

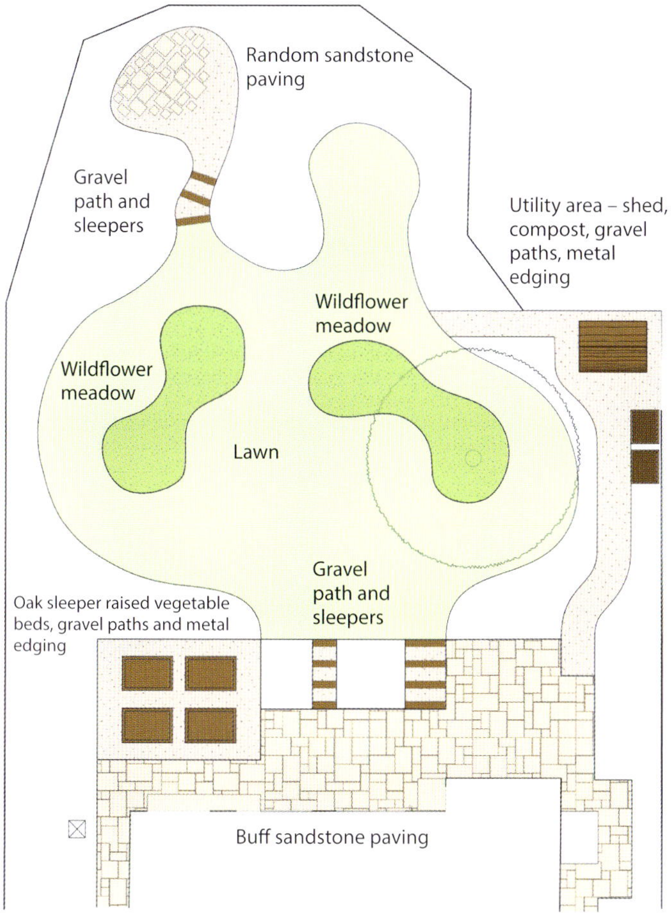

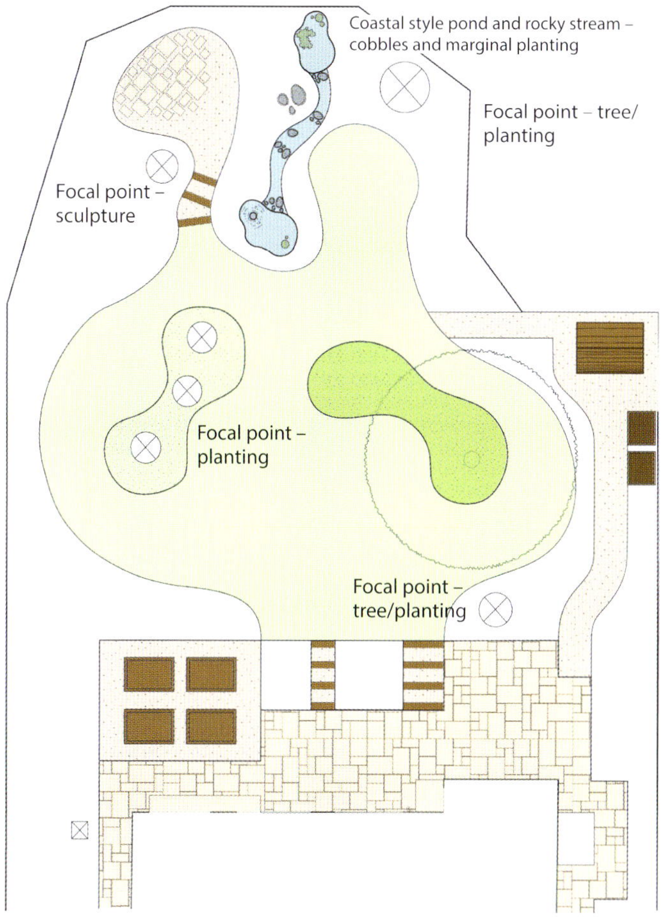

Step 5: Materials

We research which hard landscaping materials we would like to use, reflecting materials used in the house or surrounding area.

Step 6: Focal points and styling details

We think about the style effects we want to create and where we want to use planting or sculpture to attract the eye. We decide on a rocky stream feature flowing down the gentle slope at the end of the garden, a sculpture, a specimen tree that will also be part of a key windbreak, and some positions for shrubs and trees to organise views and attract attention.

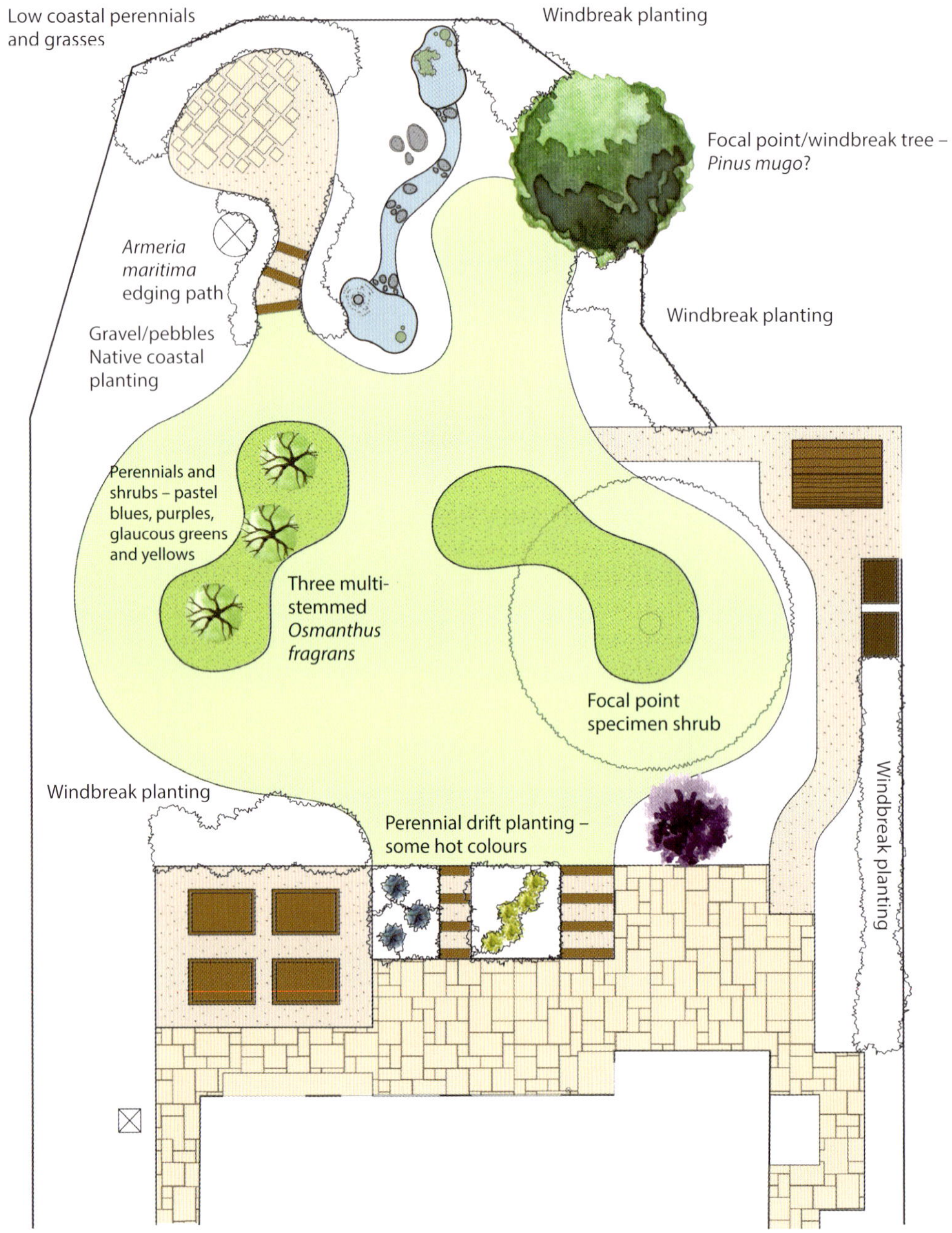

Step 7: Planting styles

We think about the types of planting we want, colour schemes, important tree positions, and shelter belt planting to filter the wind. No need for much detail of particular plants yet, but we note down some initial ideas, such as edging the gravel path that leads to our key view of the sea with thrift (*Armeria maritima*).

Step 8: Finished outline plan

We transfer all these ideas to a finished outline plan.

work and materials. If you are not confident about the details of construction methods you could leave a lot of those decisions to a trusted landscaper, but if you are at all fussy about how the features are going to look, you should do at least some research into materials and methods. If you are working as a professional, you should usually be producing these drawings. It is part of the service you are providing.

Positioning something like a patio and making sure it is the correct size is pretty straightforward when you are drawing the design but not necessarily so when it comes to actually marking it out in the garden. You need to think about what measurements are needed and where they are going to be measured from. Measurements taken at right angles to a fence line or the house are tempting but it is not always easy to judge 90 degrees on the ground. A builder's folding square can be useful to make sure your tape measure is at right angles to a boundary, but it doesn't work so well over longer distances. In those cases, it will be better to show on the layout plan how to triangulate a particular point, perhaps

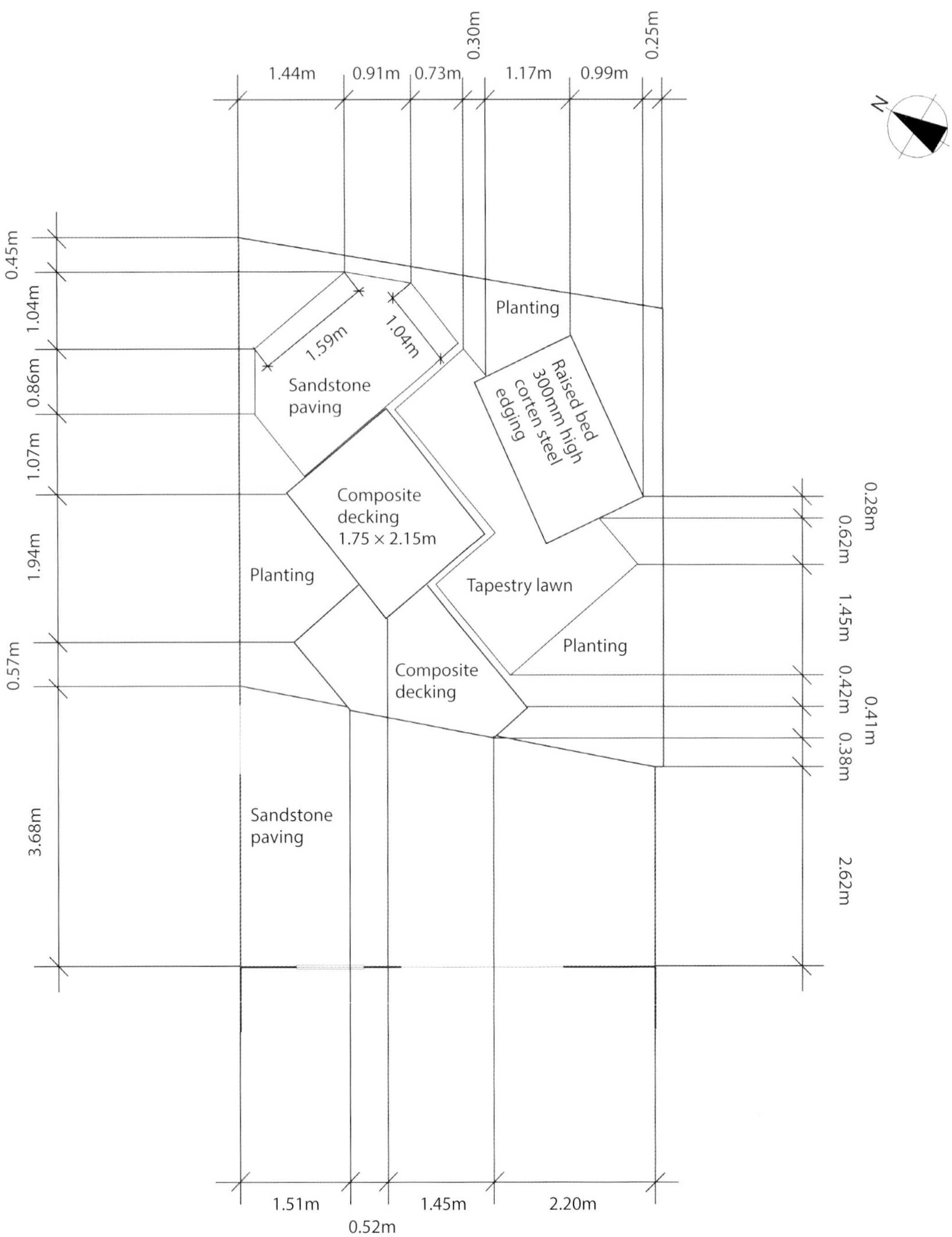

This setting out plan has measurements from known points to help set out the design on the ground ready for construction.

using two widely separated corners of the house and measuring to the point from those. For a simple rectangular feature such as a patio you might only need to fix the position of two corners. More complicated shapes will need more points indicated. Circular features such as a lawn will need the position of the centre and the radius indicated. A complex curved edge planting bed would probably be best indicated with a series of offsets measured from a boundary.

The layout plan should also show the finished levels of features such as patios, steps, lawns, and anything that is going to change in level during the build. Your existing level measurements would also be useful, particularly for preparing estimates of quantities of soil to be disposed of or brought in when, for instance, you are changing a sloping garden to a series of terraces. Be careful that the two sets of levels, existing and final, are clearly differentiated, perhaps in different colours.

The amount of detail you go into on the layout plan depends on whether you are building it yourself, or what your relationship and confidence level with your landscaper is. You need to strike a balance between having enough information to accurately lay out the design and not overwhelming the plan with information until it becomes unreadable. A larger, more complex design should probably be separated into two or more layout plans, perhaps one for the shape and position of the main features, and one for levels and extra details such as drainage or power cables.

If you are going to be working on your own garden and will be doing all the hard landscaping yourself then you might not need any formal construction drawings at all. If you are confident in using a quick pencil sketch to work out the quantities of materials that you need, then that is fine. If you are intending to employ a landscaper and don't know much about construction techniques then you could just rely on the landscaper's expertise to interpret your outline plan, advise you on the choice of materials such as paving stones, and construct it as they see fit. Sometimes suppliers of landscaping materials such as paving stones or pergola kits will supply construction drawings or have them on their website. But at least some research into how the features on your plan could be built is going to be useful. It will help you in your discussions with the landscapers, making sure that they are going to achieve the effect you want and build it to a decent standard. If you are working professionally, you should be going into more detail about the construction. If you have a good and long-standing relationship with a landscaper you probably won't need to go into every detail of what is to be done, but anything a little unusual or any detail of style or construction that may not be clear from the outline plan should be put on a construction drawing.

Construction drawings are a larger, more detailed view of particular parts of the design. The scale is usually 1:10 or 1:20. So far you have been working in plan view, as if looking down on the garden from directly above. This view can also be useful as a construction drawing, going into greater details about the exact layout of things such as the paving stones and edging blocks for a patio. But it is also very useful to produce cross sections, which are views of a feature as if it has been cut through to show the different elements of its construction. These are the main types of construction drawings that garden designers use, but it can also be useful sometimes to produce three-dimensional or isometric drawings to show how things such as the timbers of a pergola might fit together. These are a little harder to draw and if you want to work as a professional designer, it would be a good idea to get some training, but design software such as SketchUp can make drawing in three dimensions fairly straightforward. Each construction drawing and cross section should have an alphanumeric reference code that is also indicated on the setting out

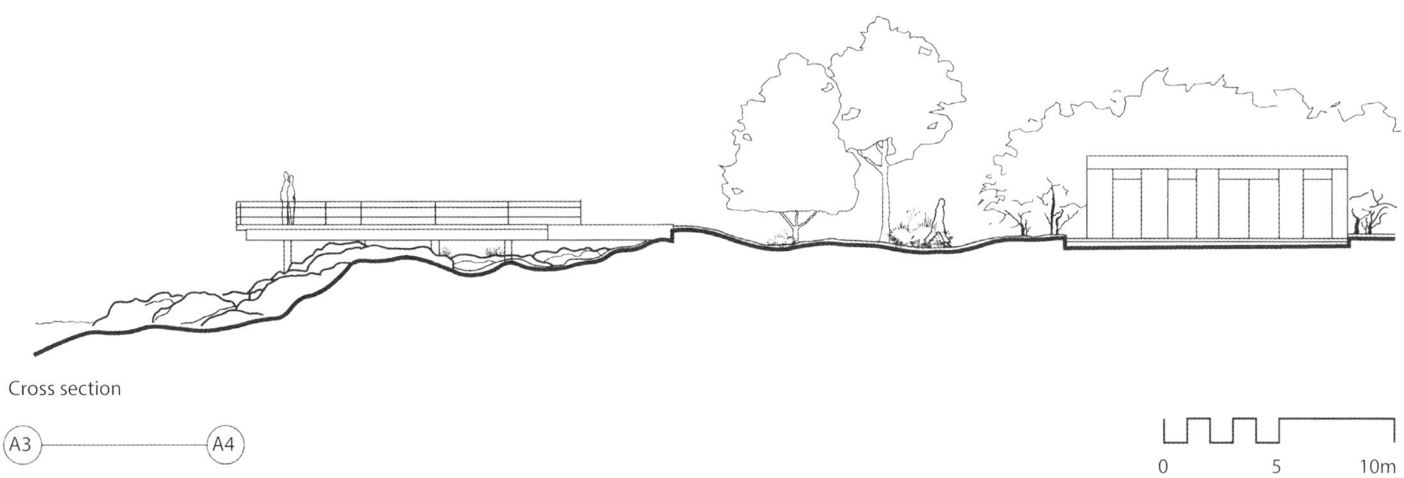

This scale drawing shows the vertical relationships of elements along a line cut through the design to produce a cross-section.

plan to show which drawing refers to which feature on the design.

There are examples of detailed construction drawings for individual features in Chapter 3: Hard Landscaping. The types of materials to be used are labelled and shading patterns used to make the different materials stand out from each other. Dimensions are labelled with leading lines perpendicular to the part and a measurement, usually in millimetres. Clarity is very important in these drawings. If you are working professionally, they are for the use of the contractor, so they don't need to be attractive drawings, just very clear. It can be helpful to use a different colour for the dimensions but other than that, colour is not often used. Even if you are not working professionally these drawings can help you think about issues that are not always clear in plan view, such as how the gravel surface of path is going to be kept off the surrounding lawn, or how earth is going to be retained when there is a terraced change in level. And, even if you don't know the exact details about how a feature is going to be constructed, a drawing, even if it is just a rough sketch, can still be a very useful way of communicating your idea to a landscaper.

6. The planting plan

If you love gardens, and we assume that you do if you are going to all this trouble to design one, this is probably the most exciting part of the design process. All of your work so far has been concerned with producing an outdoor space that works practically, has an aesthetically pleasing layout, uses stylish landscaping materials, deals with the particular problems of a coastal setting, and makes the most of the available views. But it is the planting that usually contributes the most in terms of excitement, atmosphere and satisfaction to the finished garden. The planting can also help deal with problems particular to a coastal setting such as salt-laden winds. The right choice of planting can do a great deal to improve the environmental impact of the garden by providing wildlife habitats or reducing the need for harmful inputs, such as pesticides or the fuel needed to run a lawnmower. The changing climate means that we should also give some thought to how our choice of planting could reduce the amount of water we need to use to maintain it. Chapter 6 will go into more detail on these issues.

How we choose the right plants for a garden and how we combine those plants to create a beautiful and successful garden are at the heart of what we do as garden designers and can be a complicated matter. We have broken this process down into a step-by-step process, hopefully fairly straightforward even for novice gardeners, in Chapter 4. Chapter 5 then goes into detail about the plants that perform well in coastal gardens and includes some sample planting schemes. In this section we will just concentrate on how your planting plan should be presented on paper. These should be clear, working documents that you can use on site to layout the plants ready for planting. If you are working professionally you will also use these drawings to communicate your planting ideas to your client. It is also helpful to supply your clients with photographs of the plants you are suggesting. There are various ways of sharing a photo album of plants online or you could paste the photographs onto a style sheet document and print it out to give to less technically able clients.

Planting plans are usually at a scale of 1:50 or 1:20. The larger scale is useful if you have a complicated, detailed planting with lots of different, smaller perennials. This usually means that you will need several sheets with different planting beds on each sheet.

Plants are drawn as circles with their centre point marked. The circles should be of scaled diameter that represents the plant's eventual, mature size. If you or your client is a bit impatient and you want a more instant effect, you could draw the plants at two thirds of their mature size and so pack them in a bit. It is not best horticultural practice to plant this way as you will eventually need to thin the planting out as they overcrowd each other, but it is an option.

Plants should be labelled with the Latin name. Smaller shrubs and perennials are usually planted in groups, and you should represent this by merging all the circles into one shape. The group is labelled with the number of plants, e.g., 7 – *Erigeron glaucus*. Existing plants already in the garden should be on the plan but clearly differentiated, in a different colour and with 'Ex.' after the plant name. It can also be helpful to differentiate trees and shrubs from the smaller perennials as well. You could do this by using different colours or line weights (thickness of the line) to draw the circles.

Once all the planting plans are finished you should produce a planting schedule that lists all the plants used and the size you will plant them at. The list should be alphabetical and have different sections for trees, shrubs, perennials and any other categories such as bulbs, annuals or wildflower plugs. This list is used to order the plants from the nursery and a copy should go to the client as well, if you have one, so that they know what you will be supplying.

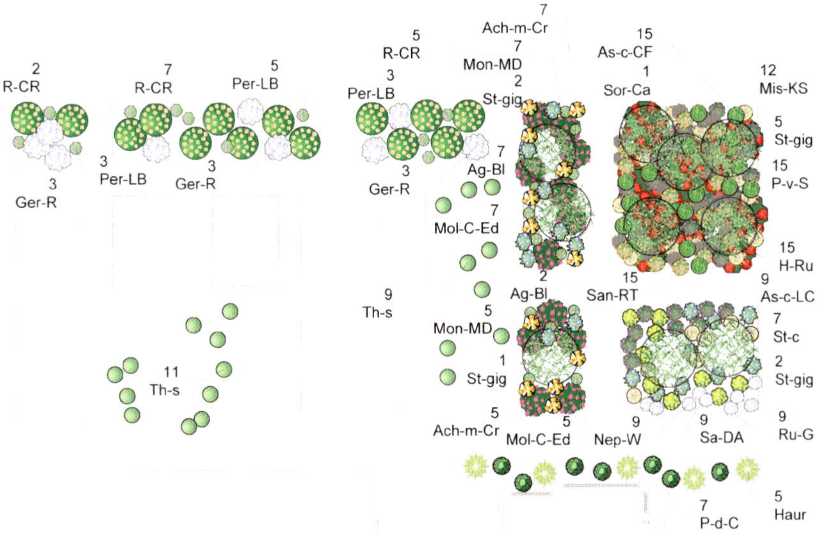

1 Planting Section A
Scale: 1:100

Quantity	ID	Latin Name
12	Ach-m-Cr	Achillea millefolium 'Credo"
9	Ag-Bl	Agastache 'Blackadder'
15	As-c-CF	Aster coelestis 'Coombe Fishacre'
9	As-c-LC	Aster cordifolius 'Little Carlow'
9	Ger-R	Geranium Rozanne
5	Haur	Hosta fortunei 'Aureomarginata'
15	H-Ru	Helenium 'Rubinzwerg'
12	Mis-KS	Miscanthus sinensis 'Kleine Silberspinne'
12	Mol-C-Ed	Molinia caerulea 'Edith Dudszus'
12	Mon-MD	Monarda 'Marshall's Delight'
9	Nep-W	Nepeta subsessilis 'Washfield'
7	P-d-C	Phlox divaricata 'Chattahoochee'
11	Per-LB	Perovskia Lacey Blue
15	P-v-S	Panicum virgatum 'Squaw'
14	R-CR	Rosa Crocus Rose
9	Ru-G	Rudbeckia fulgida var. sullivantii 'Goldsturm'
9	Sa-DA	Salvia superba 'Dear Anja'
15	San-RT	Sanguisorba officinalis 'Red Thunder'
1	Sor-Ca	Sorbus cashmiriana
7	St-c	Stipa calamagrostis
10	St-gig	Stipa gigantea
20	Th-s	Thymus serpyllum

Bowercot GARDEN DESIGN

Project:

Client:

Sheet Title: Planting Section A 2 of 7

Designed By: Alan and Bryn Edmondson

Scale: **Filename:**

Rev: **Date:** 20.3.19

The planting design shows details of the intended types and positions of all the plants in a planted area.

Once you have produced the drawings outlined in this chapter you should be very well prepared to commence the practical work of creating an exciting, beautiful garden that answers your, or your client's, brief. You should also set up a file to keep all other documentation relating to the project, such as quotations from landscapers and order confirmations from plant nurseries. This is particularly important if you are working professionally. It should then also include things such as specifications for the hard and soft landscaping, health and safety documentation, details of tree protection orders and other important planning considerations, and all contract documentation. Preparing these is beyond the scope of this book but a good course of education in professional garden design should cover it. Alternatively, you should be able to receive useful guidance from your country's professional organisation for garden designers, if it has one. In the United Kingdom, the Society of Garden Designers is very useful in this respect and is well worth joining.

Chapter 3

Hard Landscaping

When you first begin to imagine your ideal garden, you might be conjuring images of drifts of pastel flowers under gnarled ancient olive trees, but it is just as likely that your first thoughts are of entertaining on a beautiful patio on a warm summer's evening, or listening to the gentle sound of water trickling over rocks whilst sat in a secluded arbour. And it is unusual, a luxury almost, for a garden to need only be beautiful; for most people their garden has important practical functions that must be considered before it is filled with attractive planting.

If you are working professionally as a garden designer, or aspire to be, it is likely that you are drawn to the work through a love of plants. Some very good designers come from other related professions such as architecture, and they may see gardens more as constructed spaces with plants playing more of a supportive role rather than being the main attraction. Landscape architects are often as excited by the possibilities of, say, a rammed earth wall, as they are by a new variety of hardy geranium, and as garden designers we should learn from and be influenced by this approach. But for most people in the profession, like Tolkien's hobbits, our hearts lie in our shared love of things that grow. For most gardens, a large proportion of the money spent creating them will be allocated to hard materials and their associated labour costs. We owe it to our clients to make sure this money is spent on well-designed hard landscaping features that are attractive, long lasting, and that fulfil the function required of them. And if nothing else, the pictures we make with our beautiful planting schemes deserve a beautiful frame.

COASTAL STYLE

Although a coastal garden might include any of the typical hardscaping features that a garden anywhere might have, some features or styles are particularly suited to a coastal setting. As with the plants we choose, it is a good idea to take inspiration from the materials found in the landscape around the garden. We will discuss some practical considerations about materials in a coastal setting in the next section but let us first think about the aesthetics of hard materials in a coastal garden.

Pebbles

Pebbles are not good to walk on, but they do create interest and a variation in texture in a gravel garden and evoke the feeling of a pebble beach. You might scatter them randomly in the gravel, perhaps mixed with larger rocks and low-growing plants such as creeping thyme (*Thymus serpyllum*). But they tend to look better with a bit more form to their placement, swirling patterns of cobbles in the gravel subtly suggest the movement of the sea whilst still looking quite natural.

Driftwood and other beach finds can be combined with a careful choice of hard landscaping materials to create a garden even in the most exposed situations.

WHAT IS HARD LANDSCAPING?

Hard landscaping or 'hardscape' is an industry term that refers to elements of the garden that are constructed rather than planted. In general, the materials used will be inanimate, artificial, or natural, such as gravel, paving stones, rocks, timber, concrete or bricks. Although they will weather and degrade over time (sometimes, although not usually, in pleasing ways) they are primarily designed as solid, unchanging features. This contrasts with the soft landscaping which refers to all the plant materials used, materials that by their nature change over a yearly cycle and over their lifetimes.

Hard landscaping mostly involves building materials, but it does also include site grading and drainage, the shaping and moving of earth in the garden to create such things as a level lawn, or to collect and direct rainwater. Other hard landscaping features include patios, decks, built barbecues and fire pits, ponds and water features, irrigation systems, lighting, retaining walls, and raised beds.

Often professional garden designers will hand over the hard landscaping side of things to a separate business, usually a landscaper, and then return once the construction is finished to complete the planting of the garden themselves. Some businesses offer a complete design, building and planting service, offering both hard and soft landscaping, but many garden designers avoid this side of things and probably shouldn't be trusted with a disc cutter anyway. Hopefully, we make up for this with our design flair and in-depth knowledge of plants. But it is very important that a designer knows how to design hard landscaping features and understands their construction, even if they lack the practical skills. We will provide examples in this chapter, particularly as it applies to coastal environments, and if you want to work professionally as a garden designer you should be sure not to neglect this part of your design knowledge. Or, if you are employing a designer, it is worth checking, perhaps by looking at some construction drawings in their portfolio, that they are skilled in designing hard landscaping as well as planting schemes.

Beautiful Florida oolite stone is used on this garden by Craig Reynolds Landscape Architecture.

Crushed seashell

A very effective, coastally inspired, material that you can walk on is crushed seashell. A traditional path material in some coastal areas, their sharp edges are not ideal for bare feet but are a comfortable walking surface otherwise. A winding narrow path of crushed shells, edged with a coastal planting such as Thrift (*Armeria maritima*), leading the eye towards the ocean could be a very atmospheric feature in a coastal garden. They can also be used as a mulch on planting areas or containers. Non-crushed, whole shells are also available and could be used along with cobbles in a gravel garden.

Pebbles, cobbles and rock with *Agave americana* planted on a very exposed sea front.

Careful choice of stone creates a wonderful warm and harmonious effect for this Mediterranean terrace in a garden designed by Julie Toll.

Local stone

A good first source of possible inspiration for a lot of garden design choices is to look to what occurs naturally in the area. In coastal areas, the underlying geology is often more visible than elsewhere. If the local stone is quarried and available as paving stones, there does seem to be something harmonious about using it. In the United Kingdom some possibilities are:

- the famous Purbeck and Portland limestones from the Dorset Jurassic Coast;
- the gritstone sandstones, mostly associated with the north of England, including Yorkstone;
- slate, typically associated with North Wales and Cumbria but also including the sea stacks and cliffs of South Devon and parts of Scotland and the West Coast of Ireland;
- the granite deposits very much associated with Scotland but also found in Devon, Cornwall, Cumbria and in some small areas of Wales.

Grey porcelain paving, white rendered walls, and black stained timber detailing.

Paving stones from these quarries are likely to be among the more expensive options but they are often of high quality. They also may be a more sustainable and environmentally friendly option than stone imported from overseas, particularly if they are reclaimed and reused, which is often possible with something as hard wearing as Yorkstone.

Even if the budget doesn't stretch to a high-quality local stone, it might be worth choosing an imported stone that works harmoniously with the location. Much of the sandstone paving available comes in warm shades which suggest the beach. Very light coloured, reflective, almost white paving, such as some of the vitrified porcelain products, suggests a fresh, clean, sun-drenched seaside feeling. But the sun can be quite intense in many coastal settings and these very light-coloured paving stones might be a bit dazzling if used in quantity over a large patio. Bright, hot, seaside sun might also be a good reason to avoid very dark grey or black paving, as it will tend to absorb a lot of heat during the day to the point that it is uncomfortable to walk on.

Colour schemes

The most common colour scheme associated with the seaside is probably blue and white, the colour of a clear summer sky over the sea, the sun reflecting off the foam of crashing waves or the dazzling white buildings of a Greek island village. A small, enclosed coastal garden is certainly very effective with white rendered walls, raised beds or built-in seating, perhaps with blue detailing in the choice of pots, soft furnishing, or windowsills.

White rendered raised beds are a look that has perhaps had its day in the wider world of garden design. It was everywhere a decade ago but can look a bit unimaginative now. If any garden calls for it, though, it is the coastal garden. We can add a bit of interest to bring the look up to date, perhaps for one section of raised bed the rendered wall could be painted dark grey, black, or even a bold, bright colour for some contrast. Or perhaps we use a slatted trellis with a natural wood stain as a colour and texture contrast.

Black can also be a coastal colour, particularly for timber, evoking traditional fishermen's huts. Derek Jarman's cottage and garden on the beach at Dungeness is a good of example of this look. The pitch-black timber walls for the cottage with bright yellow window frames provide a very effective backdrop and complement the planting.

Shou sugi ban, a traditional Japanese technique of weather proofing that involves charring a cedar wood surface, produces some deep charcoal black timber colours with interesting textures. It has been quite on trend in recent years in interior and exterior architectural design and is just beginning to be seen in garden design. It was traditionally used as a way of protecting homes from salt ocean spray along Japan's coastline, so it could be a very useful and interesting material to use in a coastal garden.

The bright, intense light we often find in coastal settings should also encourage us to use bright, vivid colours in our garden designs. Certainly, more muted colour schemes that would be subtle and tasteful in a country or urban garden can be drained of all impact in a seaside setting. Unless you are feeling very bold, less is probably more with these colours. They are perhaps best used as accents:

Designer Rose McMonigall used a tradition of shell cladding, local to coastal regions of north-west Spain, for the wall in this show garden at RHS Hampton Court Show.

Derek Jarman's very effective use of objects found amongst the shingle surrounding his cottage at Dungeness.

some rich orange outdoor cushions, for instance, that remind us of sun setting over the sea; or a raised bed painted cobalt blue, evoking the bold colours used by Mexican architect Luis Barragán.

Set dressing

For a few final touches, it is tempting to do a bit of seaside 'set dressing' and it can be very effective, but, as with many good ideas in design, a bit of restraint and sometimes avoiding the more obvious ideas that first come to mind can be the best way. A garden full of lobster pots, scallop shells, and assorted nautical antiques is in danger of looking a bit kitsch. A derelict old boat used as a planter is quite a tempting idea but maybe it has been done too often. Of course, you should feel totally free to use any of these things if you are creating your own garden. The only person it needs to please is you, and nothing is more depressing than a garden designed just to be tasteful and inoffensive to others rather than to inspire genuine personal joy. A bit of beachcombing could certainly turn up some interesting things, perhaps some rusty old iron to use as plant supports as Derek Jarman did, or maybe an attractive fossil-laden rock. But be aware of local regulations. In the United Kingdom, the removal of natural materials from public beaches is illegal under the Coast Protection Act 1949. Recycled containers as plant pots are another possibility (perhaps old olive tins as a reminder of holidays in the Mediterranean), as is a sail canvas used as a shade canopy over a patio or deck.

A sail canvas shade is an atmospheric and practical addition to this pergola in a garden designed by Lynne Marcus.

MATERIALS

As well as being a challenging environment for plants, coastal conditions can be hard on the strongest of building materials. Salt and the increased chance of strong, stormy winds mean that we can expect most materials to deteriorate at a much faster rate in an exposed coastal garden. Being aware of the prevailing wind, knowing where the more sheltered parts of the garden are, and creating or making use of existing planted shelter belts or windbreaks will help. And we can be careful in the way we design hard landscaping features and choose materials that are best suited to cope with the conditions. In this section we look at some of the options.

Timber

Wood is a good choice of material for many garden projects such as decks or pergolas. It has a high strength-to-weight ratio and is easy to work with. But coastal conditions can be especially hard on timber. Intense sunshine will not only dry timber out, creating cracks, but the ultraviolet radiation will also tend to degrade and crack any paints or sealants used to protect it. These cracks will allow salt water to seep in, helped by the driving winds typical of many coastal areas. When this water dries, salt crystals form in the cells of the wood, pushing the wood fibres apart. Added to this, the sheer physical stress put on wooden garden structures by coastal gusting winds means that weaknesses in the materials or construction techniques are soon tested. To ameliorate these effects and prepare for them we can choose natural resistant woods or composite materials, use paints or sealants designed for coastal exterior conditions where appropriate, and use good construction techniques to limit moisture absorption.

In general, hardwoods are a better option than softwoods, being more durable and absorbing less moisture. Cedar and redwood are softwoods but have a natural moisture resistance, which means they may be suitable for some projects, although they do need to be maintained with paints or stains suitable for coastal climates. Tropical hardwoods such as ipe have been very popular and usually don't need treating with preservatives, although oil can be used to help maintain their natural colours. Ekki and greenheart are tropical hardwoods that are popular in larger scale marine construction projects such as boardwalks and groynes, but they may also be suitable for some garden projects and are very strong and durable in coastal settings. Tropical hardwoods do have the reputation of coming from unsustainable sources and being associated with deforestation; this does not need to be the case however, and the easiest way to check that it comes from a reputable source is to look for FSC certification. FSC stands for the Forest Stewardship Council, an organisation that manages a worldwide certification that promotes the highest environmental standards for sustainably managed timber.

Hardwoods can be expensive, however, and pressure-treated softwoods are still a possibility in a coastal environment. They do need a suitable treatment level depending on the amount of salt water they will be exposed to. A good supplier should be able to talk you through the options for your project.

Increasingly though, as the materials have improved in quality and appearance, composite products, made from recycled plastic and sawdust, are the most practical, cost effective and durable option. They are particularly used for decking projects but are also available in kit form for pergola construction. Strong and durable, they require no painting or staining and are very well suited to coastal garden projects as salt water has a difficult time

Sleepers, set on their end, used to create a raised planting that is reminiscent of beach groynes and drifts of shingle.

Designer Sue Adcock uses timber as a very effective material for path construction in this coastal setting.

penetrating them. Good quality products are also strongly resistant to ultraviolet light.

When designing wooden structures for coastal gardens, extra care also needs to be taken with the points of contact between pieces of timber. These points tend to be places where moisture will build up, so it is better to use, where possible, a single large timber rather than two smaller pieces doubled up. Where pieces must be joined, the area of contact should be as small as is practical and suitable spacers should be used to allow for air circulation. Horizontal surfaces where water may gather, such as the top of fences or pergolas, should be kept as narrow as possible, or capped with a sloped top surface to encourage runoff. Stainless steel screws or fasteners should be of a type that is suitable for the conditions; it is possible to buy products that have been specifically engineered to withstand the corrosive conditions found in coastal gardens.

Metals

Metals, arguably, have an even tougher time in exposed coastal conditions than wood. The normal rusting process of steel, the reaction between the iron and atmospheric oxygen to produce iron oxide, is catalysed by the presence of salt water, substantially speeding up the corrosion process. Galvanised steel, or steel treated with other metallic coatings and protective paints do not fare much better as salt- and sand-laden winds tend to strip away the protection. Weathering steels, such as Corten, which have been very on trend in garden design recently, are also best avoided. The rust patina which they are designed to have does not form well when exposed to salt water, or salt-laden winds.

Stainless steel is the most resistant to coastal conditions, but it does come in different grades with differing amounts of corrosion resistance, and different cost implications. 304 stainless steel is the most common but will stain in coastal conditions. 316 stainless steel, which has 2–3 per cent molybdenum, is the most commonly used in coastal construction but if surface finish and lack of staining are very important, then the cost of even higher grades may be justified.

The natural oxide defensive layer of aluminium is also broken down by the presence of salt. Higher strength alloys are available, and anodising or painting, along with regular cleaning may be necessary to keep it looking at its best.

Although we might not think of metals as particularly common materials in garden design, beyond perhaps some sculptures or water features or the recent popularity of Corten, it certainly has its uses. Stainless steel balustrades can give a very luxurious, coastal style to decks. And even when not being used as the main material for a garden feature, we need to be aware of what metals are being used in the details of construction, for example deck screws or other fastenings.

This concrete has deteriorated substantially in a very exposed position, but it has created interesting, perhaps pleasing textures.

Reinforced concrete can also be a problem as the rebars are usually made from unprotected steel. This isn't usually a problem as the steel bars are protected from rusting by being embedded in the concrete but in coastal conditions salt water can make its way to the steel through cracks in the concrete. The steel is then prone to corrosion and as it corrodes it expands, causing stress and deterioration to the concrete. One defence against this is to bury the steel deeper in the concrete, creating a larger gap between the outside surface and the rebars. If this is an important concern and conditions are particularly corrosive, it is also possible to buy bars that are either epoxy coated, made of stainless steel or made of fibre reinforced polymer.

Bricks and render

Bricks and concrete blocks are, in general, highly durable materials but we still need to take some care in the design of any brick or block structures in coastal gardens. The likelihood of strong gusting winds means that we should take extra care that any brick features are designed to be strong enough to deal with these extra wind loads. Detailing, such as stone caps or coping, is even more important in the construction of exposed walls in coastal areas, to protect them from damage and staining resulting from driving coastal winds and rain.

Lime mortar is more porous than cement mortars so will absorb more water. When this evaporates, any salt in the water will crystallise and damage the mortar, causing it to become powdery. However, this process wicks salt water out of and away from the actual bricks in a wall, resulting in less salt formation on brick surfaces which is both unsightly and eventually damages the brick. Cement mortars will not have this beneficial effect but will last longer than lime mortars. So, in some cases, there is an argument for using lime mortar and treating it as sacrificial and regularly replaced in order to protect the long-term appearance and integrity of the wall. But where the wall is very exposed to salt winds lime mortar would probably not be practical or recommended.

Calcium silicate bricks, common in some areas such as North Germany and Sweden, are very subject to salt damage and should not be used in exposed coastal situations.

Rendered concrete block walls are a common feature in modern garden design. As well as protecting the wall, the render gives us, as designers, a chance to create wide expanses of solid colour and neat, clean lines. Where these walls are exposed to strong coastal winds, extra care should be taken that the render will be able to cope with the wind-driven rain. Thicker coats of render may be appropriate, and some factory-produced renders are more suitable for these conditions than the standard sand and cement mix. It is also a good idea to use PVC render beads rather than galvanised or stainless steel as any rust will stain the render.

CONSTRUCTION DRAWING – GABION AND PLANTED WALLS

In this section we will look in detail at two hard landscaping features that might be useful for your coastal garden design: the gabion wall and the planted, dry-stack or dry-stone wall. These will also serve as examples of the sort of drawings you should be producing if you are working professionally. Producing construction drawings may also be useful if you are working on your own garden, either as instructions for your own later work or as a communication tool for dealing with anyone you may employ to help you.

The details in this section are intended only as a basic guide to give you some ideas of features you could include in your design. If you are not particularly skilled or knowledgeable in this area, you should work with the help of a landscaping, building or design professional. If the feature you are designing involves significant structural or safety implications, such as a large wall retaining a steep slope or area with buildings, you should involve an engineer. And if you are working professionally, you should investigate professional indemnity insurance for your business.

Gabion wall

Gabion walls are built from a series of rectangular steel baskets, containing rocks, and wired together. The baskets are usually supplied in flat pack form and assembled and filled with local stone on site. They are used extensively in large scale construction projects as retaining walls, holding back soil, and preventing slippage and erosion. Their weight and wide base make them very useful in this context, as does their permeability which allows water in the soil behind the wall to flow away, reducing the hydrostatic pressure on the wall.

But as well as their structural capabilities, gabions have much to recommend them from an aesthetic perspective for use in a coastal garden. They are quite often seen in coastal contexts, helping to ameliorate slippage and erosion, and so work as a reference and response to the environment for gardens in coastal settings. It also gives us an opportunity to use local stone, showing the variations of colour and texture to good effect. And if you are using stone as part of another feature, perhaps a beach-styled gravel garden, it continues that effect. Or it could match your choice of paving stone. It is, in general, a good idea to limit your use of different hard landscaping materials; using the same material in different contexts can make your design more coherent in a pleasing way. For these reasons, as well as functioning as retaining walls, they can be used just as freestanding features, as boundaries, or perhaps

WORKING WITH LANDSCAPERS

If you are working professionally, a good course in garden design should provide you with the skills to design and produce specifications for most common garden features. You should also, unless you are planning to offer a complete design and build service yourself, develop relationships with some good landscaping firms local to you. Their experience of the on-the-ground practicalities of the construction side of making a garden is going to be deeper than yours, particularly early in your career, and you should be open to their input. For instance, they may suggest that you have over specified on some of the materials and that costs can be saved without compromising significantly on structural integrity or design effect. Be aware, however, that they have their own agenda, which may not exactly align with yours or your client's. Do not be afraid to argue your position if you feel any changes will spoil the design in some way. This is something that becomes easier with experience, but your first duty is always to your client, and you must endeavour to deliver a garden that lives up to the promise of your design. Taking the time to produce a construction drawing and specifications for the construction and materials makes it clear from the start what you are aiming for. It also means that if you arrange to have several landscaping firms produce quotations for the work, they should be quoting for the same thing and so the quotations will be easily comparable. The details may change and evolve during the construction process, but you should make sure that you are part of the conversation if any major changes happen, and those changes should be documented so that there are no misunderstandings later.

These substantial gabions, retaining a steep bank, would need expert input with their design and construction but smaller versions are a feasible DIY project and work well in coastal design.

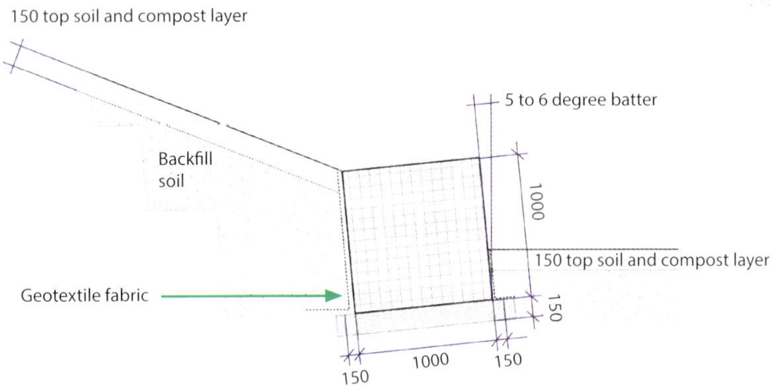

Construction drawing for a retaining gabion wall.

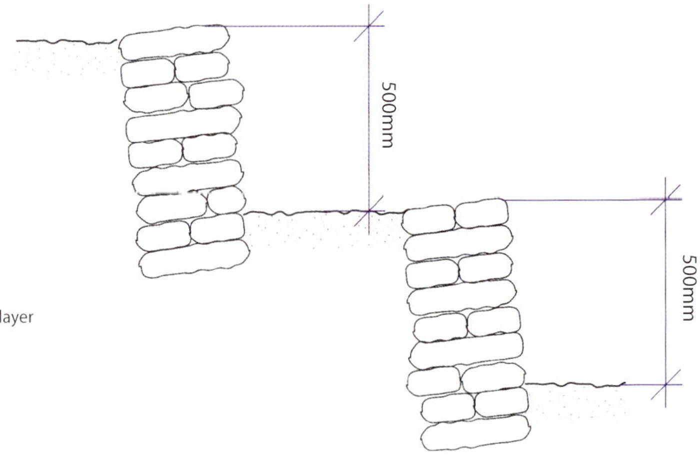

Two spaced, smaller planted walls are usually more stable than one higher wall.

planted. It is also straightforward to create planting pockets in the wall, such as native sedums.

Smaller gabion walls, particularly freestanding ones less than one metre high, are fairly easy to build, and may be achievable as a do-it-yourself project, but they are heavy once filled and care should be taken, particularly if working on a slope. For larger walls, you should get professional help, and for those required to retain steep slopes, or slopes with buildings at the higher level, need an engineer to help with the design.

In a coastal garden, exposed to salt-laden winds, it is important that the cage is made of a steel that is suitable for these conditions. Stainless steel and PVC coated gabion cages are available and you should check with the manufacturer that they will have a decent corrosion resistance and lifespan in a coastal environment. The manufacturer should also provide advice on assembly and construction. The details and drawings below show how a wall of 1 metre high or less, freestanding or retaining a stable, flat area could be designed. Such a wall should have a height to thickness ratio of no less than 2:1. Walls that are narrower than this may not be able to cope with the lateral force of the retained soil and could overturn. If you have any doubts, you should seek professional advice.

A 1m (3.3 feet) high gabion wall should not need a concrete base and can be set on a 50–75mm (2–3 inch) layer of compacted crushed carboniferous limestone, known in the United Kingdom as MOT Type 1, or scalps. If it is a retaining wall, a geotextile, membrane fabric behind the wall will prevent fine soil from running into the gaps between the stones. Over time this soil could reduce the free draining capability of the wall and increase the hydrostatic pressure acting on the wall. Geotextile fabric can also be used underneath the wall to discourage weeds from growing up through the wall, although the layer of compacted MOT Type 1 may be sufficient to prevent this. A retaining wall should also be built at an angle, or batter, leaning back into the retained soil.

It is also good practice to include a layer of drainage gravel and a perforated drainage pipe behind the wall. The cages should be firmly wired together with lacing wire, helicals or 'C' clips, and have internal bracing ties to prevent deformation and bulging. The manufacturer should be able to provide these.

Some care should be taken with the choice of stone to fill the gabion baskets. The pieces must not be smaller than the mesh and should not be susceptible to frost damage. It is important to prevent voids when the stones are packed into the cages and this is easier, and the stones interlock better, if they are angular pieces rather than rounded. Voids or stones that have been degraded by frost damage will collapse over time and this may cause the wall to fail.

Planted wall

An alternative to the gabion wall as a retaining or freestanding feature is the dry-stack or dry-stone wall, that is a wall composed of rocks stacked on top of rocks without the use of mortar. It is like the gabion wall in that it is a gravity wall, held in place by just the weight of the materials used, and that it aesthetically allows us to make use of the colours and textures of natural stone and its coastal associations. Also, like the gabion wall, it can be planted and is even more suited to this than the gabion wall. This is the style of wall we will look at in detail here. Whereas the gabion wall has a more stark, raw, constructed feel to it, the planted wall looks more natural, reminiscent of a cliff face or rock outcrop, and more traditional, being essentially a vertical version of the rock garden. It is not a building method that should be used for any

Prostrate rosemary, *Rosmarinus officinalis* 'Prostratus', growing over a harbour wall edge.

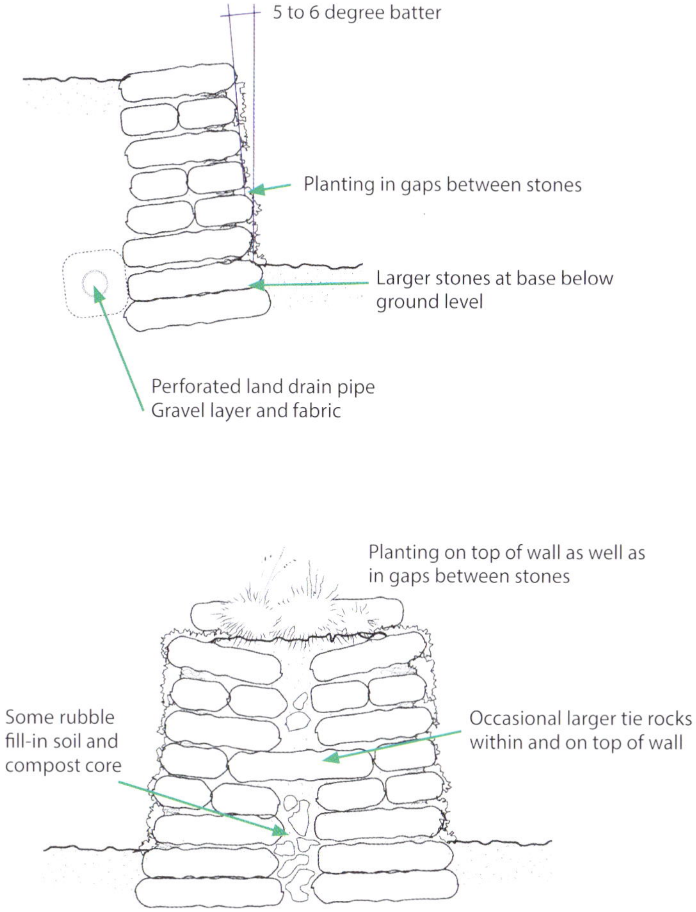

Retaining and Freestanding – Construction drawings for retaining and freestanding planted wall.

height much more than 1m (3.3 feet) because the materials are very heavy. It would be much better in terms of stability to design two walls of 50cm (1.6 feet), separated by a flat area, rather than one wall of 1m (3.3 feet).

Construction should start with an excavated, level area for the footings. If the bottom of this excavated area is at all soft, a layer of compacted, crushed limestone, MOT Type1, should be added. Then the first foundation course of larger stones should be laid below ground level. As we will be planting more extensively into the wall than we would into the gabion wall, it is probably best that we do not use a geotextile fabric and layer of drainage gravel behind the wall. We want the plants to send their roots into the soil behind the wall and this will help to stabilise the soil. But there should be a layer of gravel (and a perforated drainage pipe if the drainage is poor) at the base of the wall to prevent movement under the wall from destabilising it. The face of the wall is canted back with a batter of about 9 inches to 1 inch (22.9cm to 2.5cm) and joints should be staggered from one course to the next.

Larger rocks are more durable and stable but if you are making a wall as a do-it-yourself project there is a limit on the size of rock that you can safely and practically move. The appearance of the rocks is a very important part of the feature, so they should have interesting variations in colour and texture and perhaps pick up on materials used elsewhere in the garden or occurring naturally locally. Hard stones such as granite are a possibility but it is more usual to use some sort of sandstone, limestone, or slate. Angular, sharp edged pieces with flat surfaces are the easiest to stack and lock together but, as we do so, we should find that there will be natural small gaps in the wall. These will be our planting pockets.

The crevices and holes we leave in the wall will create very sheltered and favourable planting areas for the right plants, well drained and sheltered. In a coastal garden, any planting we do on the leeward side of a wall, facing away from the prevailing salt-laden winds, will be one of the most sheltered planting opportunities we will have in the garden, and so our choice of plants is wider than it might be for the rest of the garden. Our choice of plants is likely to come from those small, spreading plants that grow in rock crevices in their natural environment. They will often naturally trail, creep, or form carpets or small mounds. Flowering plants trailing down a sunny wall are particularly pleasing. Creeping thyme (*Thymus serpyllum*) and prostrate rosemary (*Rosmarinus officinalis 'Prostratus'*) are good bets, as are many of the sedums and sempervivums. *Armeria maritima* and *Centranthus ruber* would work well as part of a coastal planting scheme. Perennials such as *Arabis causica* 'Rosabella', *Campanula cochlearifolia*, and *Convolvulus sabatius* are reliable choices for lots of creeping or trailing flowers. Plants that would work in the cooler, shadier side of a wall would include *Armeria balearica*, *Saxifraga umbrosa* and ferns such as *Asplenium trichomanes*.

Chapter 4

Planting Design

As with the more general outline design, it is perfectly possible to create a beautifully planted garden without producing a plan – if you are prepared to patiently tweak and experiment over several years, scrapping what doesn't work and trying out new ideas until the garden reaches a fantastic balance between your sense of what is beautiful and the practical conditions you must deal with. But if you are working professionally, your client is likely not to want to wait that long. Or, if you are working on your own garden, you yourself may very well not want to wait that long. You may also want to minimise the amount of money you waste on plants that are unsuitable for the position you want to plant them in. Or you might have a particular visual effect that you want to attain. A good planting design can help you achieve all these things, right from the start, and in this chapter, we will outline a step-by-step method for producing one to a professional standard.

If you are trying to create or improve a garden in a coastal setting, there is even more reason to think that a planting plan is a good idea. A good understanding of what parts of the garden are exposed to salt winds, what parts are particularly sheltered, and where there are spectacular sea views that should be framed and enhanced with clever planting, is greatly helped by the process of producing a drawn plan of the areas you are going to be working on. As you will see in the next section when we work through a step-by-step process for producing a planting design, the survey process and outline plan that we worked through in Chapter 2 is your starting point here. But first we will give some consideration to things we can do, practical and aesthetic, with our planting that are particularly relevant in a coastal garden.

CREATING WINDBREAKS AND SHELTER BELTS

As we have already noted, salt-laden winds are a challenge in the garden and limit the choice of plants that will thrive. The wind increases the loss of water from the plant by increasing the

Pine trees and shrub understorey shelter belt for a coastal garden on Purbeck, Dorset, UK.

Succulents such as *Echeveria* and *Sedum* cultivars are excellent choices for planting in a gabion wall. The example here is from a garden designed by Sarah Morgan.

transpiration rate and the salt in the wind causes foliage to scorch. In severe cases this can result in the death of the plant. We are very limited in our choice of plants that will survive full exposure to these conditions. But some will and some of these, particularly the trees and shrubs, can be used to create windbreaks that will redirect or filter the wind, creating sheltered areas where our choice of plants can be much wider.

Hard landscaping vs planting

Solid features such as walls and fences can increase turbulent effects on the wind, which can limit their effectiveness. There are permeable options which lessen these effects but some of these, such as polypropylene webbing, are rather unsightly, and the more attractive options, such as hazel hurdles or lath fencing, might not fit the stylistic effects that you are trying to create. Well-chosen and positioned planting is often the best option to create a pleasing visual effect and the right amount of filtering and reduction of wind effects.

Positioning

When positioning the windbreak, you need to first take account of the prevailing wind direction, which is, in the United Kingdom, as we have previously noted, often southwesterly. The consideration should be given to the effects produced by the wind bouncing off walls of the property or outbuildings as this can cause the wind to come from unexpected directions. The wind passing between buildings and other structures can create a wind tunnel and this can be where some of the most severe and unexpected exposure conditions are to be found. We have often been called in to help with gardens where planting has failed because the effect of wind tunnels created by buildings had not been anticipated. Your survey and outline plan, if you have produced them, should have already taken note of these effects, and included planted windbreaks in good positions. Well-planted shelter belts and windbreaks should significantly reduce the effect of the wind for an area reaching into the garden to about ten times the height of the trees or shrubs.

Planting density

You should be as generous with the number of trees and shrubs as space will allow. If the windbreak is too small or insubstantial the wind will find its way around the edges of the windbreak or create unhelpful eddies of wind and actually increase the wind speed on the leeward side. But the planting should filter the wind rather than be so dense as to provide a solid barrier. We want to be filtering out about 50–60 per cent of the wind to be most effective.

For windbreaks, shrubs and trees should be planted a little closer together than you would in a standard garden planting scheme. The shrubs should be planted with 30cm to 1m (1–3 feet)

SHELTER BELT OR WINDBREAK?

We use the term 'shelter belt' to describe a much more substantial planting than a windbreak. They are only suitable for larger gardens and usually consist of, going from the windward side towards the garden, some large shrubs, a row of conifers, two or three staggered rows of mixed deciduous and evergreen trees, and a final windbreak planting of large shrubs that provides shelter lower down where the tree trunks may not be filtering the wind. A mix of about 80 per cent deciduous and 20 per cent evergreen works well and looks natural.

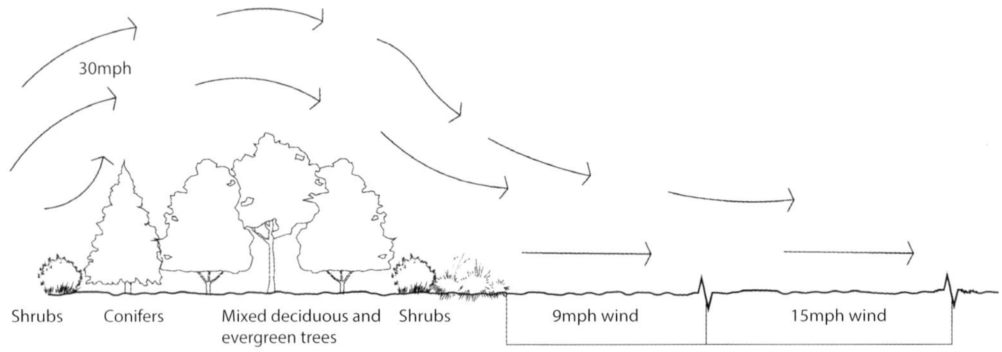

Shelter belt planting, 10 metres high.

The windbreak refers to just the large shrub/hedge part of this or perhaps one or two staggered rows of trees, or a mix of both, and can be created in almost any size of garden.

gaps, depending on the size of the shrub. The trees should be spaced about 1–2.5m (3–8 feet) apart, depending on the size of the tree, but could have an underplanting of shrubs in the gaps.

For shelter belts, the deciduous trees should be planted with gaps of about 2–3m (6.5–10 feet) between them and the staggered rows should be spaced about 1.5m (5 feet) apart. The row of large conifer trees should be about 2m (6 feet) from other rows and the trees planted with 2m (6 feet) gaps between them. Very substantial shelter belts might then have another two or three rows of smaller trees and shrubs on the windward side of the conifers, but this would only be a consideration for the largest of gardens or, more likely, commercial sites.

Cupressus macrocarpa 'Goldcrest' has been used as a windbreak in this garden, as it is very exposed to salt winds. It has suffered badly on the windward side but there is no scorching on the leeward side.

Recommended trees and shrubs for shelter belts and windbreaks

Any of the large shrubs or trees in Chapter 5 could be used to create shelter belts and windbreaks but some are more commonly used for this purpose. For shelter belts, the large conifer row might typically include *Pinus radiata* (Monterey pine), which is fast growing and very well adapted to coastal winds. Its lower trunk is bare, but the rows of shrubs and smaller trees will provide this lower storey wind filtration. *Cupressus macrocarpa* (Monterey cypress) is another excellent, fast-growing option which provides cover from top to bottom of the tree. *Pinus contorta* and *Pinus nigra* work well. *Quercus ilex* (holm oak), which is not a conifer but is large and evergreen, is also commonly used.

Any of the smaller trees, deciduous or evergreen, listed in Chapter 5 could be used for the staggered rows on the leeward side of the large conifer/evergreen row. The hawthorns and blackthorn provide a very attractive, natural effect and are an excellent wildlife habitat. Tamarisk, although usually grown as a large shrub, can also be grown as a small tree and would add a dash of colour interest.

For the large shrub row, which is the final part of the shelter belt on the leeward side, again, any of the large or medium shrubs from Chapter 5 are possibilities. You would also be selecting from these if you were creating just a windbreak for a smaller garden, although in that case you might also be adding some small or medium-sized trees to the mix. *Ulex europaeus* (gorse) is an excellent choice if you are looking for a more natural effect. *Hippophae rhamnoides* (sea buckthorn) looks very at home in a coastal garden and is easy to establish. It is worth considering using some of the larger grasses on the leeward side of the windbreak row to add a different texture and wind filtering effect, such as some of the *Cortaderias* (pampas grasses), or the larger *Panicums*.

The windbreak could also be planted as a coastal hedge and typical choices for this would be *Escallonia*, *Griselinia*, *Euonymus japonicus*, *Elaeagnus* or gorse. It could also be a mix of some or all of these. A hedge will be a more complete visual barrier so is a good idea if you are also looking to create privacy or to screen a particular view. It can also be kept carefully clipped if you are looking to create a more formal effect. Even the shrubs or trees most adapted to coastal conditions might need a bit of extra protection until they get established, particularly if they are planted as smaller specimens.

Young plants are the better choice as they establish more quickly than more mature specimens. In the average sized garden, they can be planted as a double staggered row and can be just shrubs or a mixture of shrubs and trees. If a few trees are included, shrubs should be planted between them to filter the lower part of the wind. Avoid using just evergreen shrubs, as too dense a hedge can create a barrier that blocks rather than filters the wind. Plants within the row can be planted 30–100cm (1–3 feet) apart and can be sheltered from the wind by a plastic windbreak fencing until they establish.

A pleached hedge is another windbreak option and can add a striking design element to the garden. *Carpinus betulus*, the native hornbeam, is a particularly good choice for this treatment, especially in coastal areas. A pleached hedge consists of trees grown with a clear stem with the branches trained on a bamboo frame. The hornbeam will retain its leaves during the winter but in a very exposed site may lose some in windy weather. Alternative choices include *Elaeagnus × ebbingei*, *Griselinia littoralis* and *Quercus ilex*, the holm oak. These three choices are all evergreen.

PLANTING A PLEACHED HEDGE

Young pleached trees are typically planted with a 180cm (6 feet) trunk and their branches trained on a bamboo frame of about 180cm (6 feet) square. They should be staked using two posts and a cross rail. On very windy sites, a second cross rail, halfway between ground level and the top rail, should be used. Check that the ties on the rail and bamboo frame do not become too tight as the trees grow. During the summer new shoots should be trained and tied to the bamboo frame. Lateral shoots should be intertwined and trained with those of the neighbouring trees.

When the pleached trees have grown to the required extent they can be pruned to create the hedge. Outgrowing shoots can be pruned back during autumn and winter. Once the hedge is mature, the old bamboo frame can be cut away. If the pleached hedge is on the boundary of the property, discuss with the neighbours if they are happy to prune their side of the hedge. If not, a space will need to be left for pruning access.

A pleached hedge of *Carpinus betulus* (hornbeam), three years from planting, providing shelter for the garden beyond.

As noted above, it is generally a good idea to plant smaller, younger specimens as they tend to establish themselves more quickly and adapt to the particular conditions of the position chosen for them. It also saves you quite a lot of money as plant prices tend to climb quite steeply for larger specimens. But small, young specimens are more vulnerable to the physical effects of the wind because their small root ball means that they can be buffeted about more. Because of this they may need some extra shelter in the first couple of years. It is possible to buy some hedging shrubs such as *Griselinia* as quite large plants without spending too much money. We sometimes plant a group of these behind plants that have been planted as smaller specimens. As the garden matures the *Griselinia* are no longer needed for protection but become an evergreen backdrop for the more interesting plants in front. If space is limited, some gardeners might use planting such as the *Griselinia* as sacrificial planting, digging it out once the more interesting planting has established itself. However, it does seem a shame to plant with this later destruction in mind. It is better to make such planting part of a more long-term part of your scheme if you can. Where there is not much space, plastic windbreak fencing or even artificial hedging might be a better solution for temporary screening until smaller plants get established.

Griselinia littoralis can be bought at a reasonable price for large specimens that will provide instant screening.

COASTAL PLANTING STYLE

Before we look in detail at how to produce a planting plan, it is worth considering what style of planting you are trying to achieve. If you have created a sufficiently sheltered area nearly any style of planting is possible, but some styles are more suited, practically and aesthetically, to a coastal setting. All gardens are a meeting point of the natural and wild with the domestic and constructed but a coastal garden is likely to have a close relationship to its surroundings. If you are designing a garden with any sort of sea view it is likely that that view will be the focal point of the garden and anything you do in the garden will be playing a supporting role. The extreme of this will be a garden that is right on the shoreline. It can be possible in this situation to create windbreaks and sheltered spots, but it is likely that the best approach would be to take a lot of inspiration from the natural coastal planting around the garden, subtly tweaking and enhancing it to create your planting design. A famous example of this approach is the garden that Derek Jarman created at Prospect Cottage on the shingle beach near the Dungeness nuclear power station on the Kent coast in the United Kingdom. His use of coastal wild plants such as sea kale, sea holly and gorse, along with carefully chosen non-native plants such as santolina and Californian poppies set the cottage firmly in its landscape. The bright yellow window frames, in beautiful contrast to the black shiplap walls, are echoed in the preponderance of yellow flowers. The use of a collection of driftwood and other shoreline finds as sculptural elements and the lack of a boundary defining feature such as a fence further reinforce this garden's reputation as an exemplar of how a garden can very closely respond to its environment, subtly enhancing it, but remaining very much part of its setting.

So, almost always, our first step is to look to nature. What is growing locally in similar conditions to those we are designing for? What is the look and feel of the surrounding coastline? Take a walk around the area, make notes, and take photographs. Even better, take a sketchbook and record your impressions and things you notice about the local wild plant life. You do not need to be a great artist for this to be a useful exercise. Just the act of sketching and taking notes makes you look much more closely than you would otherwise. Draw and note not just the species of plants you find but how they grow. What other plants grow with them? What effects do they create? Perhaps you notice soft grasses moving in the breeze, or the dramatic sculptural silhouette of a gnarled tree as the sun sets. Draw them and note your observations. If you struggle to identify any of the plants there are apps available for your phone that can attempt to identify anything you photograph, including plants. They are not completely reliable but are surprisingly good at suggesting possible plant species.

Although not the only way to approach planting design for coastal gardens, a more natural style of planting, using plants that, even if not actually found locally in the wild, at least look like they belong in that setting, is a very effective approach to a coastal planting. You could also take a freer, more impressionistic approach to reacting to the sense of being by or near the sea. There are more formal approaches that can work, and we will discuss this approach later in this section, but there is something about the movement of the sea that suggests flowing curves and movement. Soft ornamental grasses, such as *Panicum virgatum*, often work well for this reason.

Colour schemes

The need to reflect away intense sunlight means that the greens of coastal plants often tend toward the grey-green, glaucous shades and these colours will look more natural as part of your planting scheme palette. However, if you are painting outbuildings and fences black or dark grey, which can look fantastic in coastal gardens, more vivid greens are very much enhanced by this background. As the sunlight can be intense, more vivid flower colours are a good idea, so that they are not bleached out by the light. If you are lucky enough to have the spectacular sunset views that we often associate with being by the sea, you might reflect this in your planting scheme by including some orange and gold colours. The blues and white that we often associate with coastal interior design can also be a successful theme in your planting.

Wild coastal grasses picking up the golden colours of the sunset.

Alternative styles

The garden at Prospect Cottage is quite an extreme example of gardening in a coastal setting. Coastal gardens are not always this exposed or so close to the shoreline. The view might not always need to be so dominant, and we may look to create more formal barriers, perhaps using shelter belts or windbreaks to frame a view and create more privacy and a wider range of plant choices and planting design styles. Once we have a bit of shelter, other planting styles can be considered. The milder microclimates that can be found in some coastal settings may allow you to create a tropical planting focusing on hot colours and architectural foliage. Care must be taken to keep some of the very large-leaved plants such as bananas very well sheltered, as coastal winds will tend to shred the leaves. A good example of this sort of garden in the United Kingdom is Compton Acres near Bournemouth in Dorset.

Containers can be very useful in a coastal garden, particularly where the soil is poor. Care will be needed to keep them watered but a well-chosen selection of recycled pots and containers of all sorts, along with other gathered items of driftwood and other beach finds might form the basis of your planting design.

Similarly, if you have a sheltered enough spot, it may be possible to create a productive garden for vegetables, herbs and fruit. We will discuss later in this chapter what plants will work best for this but in general it is likely that raised beds, large planters or troughs will be the best approach to ensuring the soil is of sufficient quality.

It is unlikely that you will be creating a classic country garden style with perfect lawns or cottage garden style planting. But you may be able to create a similar feel but with a more modern and coastal-friendly approach with drifts of suitable perennials, using not one or two of each plant variety but five, seven, fifteen, twenty or more of each, depending on your scale.

THE FIVE-STAGE PLANTING DESIGN PROCESS

It is useful to have a clear, step-by-step method for producing a planting design, like the general outline design process shown in Chapter 2. Our approach to planting design works in a similar way to the outline design, starting with gathering all information about the site that you need, clarifying what you want the planting design to achieve (the brief), and finally drawing detailed plans of how you are going to use new and existing planting to answer your brief.

A very similar design process is in fact to be found in most design disciplines. A designer is likely to go through very similar steps to design a chair or an interior; it is just that the materials are different and the knowledge required changes. For planting design, you will be using your knowledge, understanding and appreciation of plants. The overall process remains broadly similar, as does the need to understand and be sensitive to colour and form, but planting design is unusual in its emphasis on changes over time. As garden designers we need to think like interior designers or architects as we work with space and colour and form, but then we need to develop our ideas almost like cinematographers as we think about how these forms and colours will work in the future.

A remarkable range of planting is possible in this sheltered part of the coastal garden at Compton Acres.

Container planting works well in this harbourside walled garden.

Different parts of the garden will have their starring moments as flowers come and go across the seasons. A quiet corner of the garden that feels sheltered and cosy when the trees and shrubs are in full leaf might become much more open and connected to the view around it when the leaves fall, and we are left with the new, more starkly architectural forms of the winter. And the garden will have a longer narrative arc, from the sparse effects of freshly planted beds to its lush maturity, to its old age when key plantings reach the end of their lives and are replaced.

This need to think of how your planting design will change over time can be tricky. Add to this the often fragile and unpredictable nature of the materials we work with, and you are left with a difficult design discipline to master. When carefully chosen plants mysteriously fail to thrive, as happens in the life of any gardener, it is tempting to give up and just tarmac over the whole garden. But there is something so deeply fascinating and rewarding about working with plants that keeps drawing us back. And the design process outlined here, combined with the plant lists in Chapter 5, should give you every chance of producing a successful and beautiful planting scheme. As we outline this process, we will assume you are working on a large design for a whole garden and will use the design we produced in Chapter 2 as an example. But a design for just a small area or single bed can be produced in the same way.

1. The schematic planting design

For the first step you will need a clean copy of your outline design for the garden. If you are using a software design package, this should simply be a matter of deleting all the labelling, new planting, shading and colours from the design, leaving you with just a line drawing of the design, and saving this as a new file. If you are drawing by hand then you can trace the outline onto a new sheet of tracing paper and work on that. Onto this we will be collecting our information and ideas and producing a version of the design which shows the on-site conditions we will be dealing with, what sort of planting we want to put where, and the functions, such as visual or wind screening, we want the planting to fulfil.

First, make sure you have any existing larger planting that you want to include in the new garden. This will just be trees or large shrubs at this stage. We can account for any smaller plants we want to keep at the detailed planting plans stage. You should have measured the positions of these plants during the survey stage of the design process in Chapter 2. Even if you are not sure that something will be kept, it can go on the plan at this stage as part of the decision-making process.

Structural planting

Now the practical requirements of the planting can be addressed. You should have identified things like undesirable views that need to be screened with tall planting, key views that should be framed and emphasised with suitable planting, and areas of the garden that need to be screened from prevailing winds with a shelter belt or windbreak. Your design should have taken these things into account, so that planting can be roughly indicated in terms of type, extent, and height on this schematic design. For instance, you could draw in where a shelter belt should go with its rough size and shape. It can just be marked as shelter belt for now; you can get into the details of this planting later. Or you might decide that a key view could be framed by two matching trees, just mark these on the plan as trees for the moment, along with an idea of the height and spread you would ideally like them to have, and again, you can worry about the exact species later.

Focal points, fragrance and other aesthetic effects

After you have placed the structural planting, you can begin to think about the planting that fulfils a more strictly aesthetic purpose. Your design should suggest places where some key focal point planting could go. Does the placement of benches or patios, combined with the shape of the planting beds, suggest spots where a particular eye-catching tree, shrub or planting combination would have the best impact? Scale is important here. If the focal point planting is to be at the other end of the garden from the main patio, then it should probably be a tree or large shrub in order to have sufficient impact. But if you have included a small, sheltered part of the garden that is to be discovered as the garden is explored, then a fantastic but small-scale planting combination of shrubs and perennials might be called for. In that circumstance you might also be thinking about fragrance. The focal point might be the fragrance of something like *Daphne odora*, encountered as you turn a corner on a path before you even see the plant itself. Mark these focal points on the plan along with any ideas you have about the type of planting you might use.

General types of planting and colour schemes

Finally, the rest of the schematic design can be filled in with more general descriptions of the type of planting you are looking for in different areas. You should have some of this already on your outline design, but you can get into a bit more detail here, including some initial ideas of the types of plants you might use and the areas they will cover. One part of the plan might be marked as 'drifts of perennials' along with some ideas of the colour scheme and a few species you might use if you have some ideas already. Another, more exposed area might be marked as 'coastal wildflowers edge the path'. Once you have a good idea of what you would like to achieve with each planted area you can proceed to detailed planting

PLANTING SIZE

A constant challenge of working on a design as a two-dimensional plan is to keep at the forefront of your mind that we are designing a three-dimensional space. In terms of planting design, it is useful to think of plants as falling into four rough categories in terms of their structure, size, and practical and aesthetic function.

Trees

Garden trees, of course, come in a wide range of sizes, from *Pinus mugo*, which might reach 10 feet or so at full maturity, to the full, soaring majesty of all 40 feet of a mature *Pinus nigra* 'Maritima'. But at all sizes, they seem to have a particular relationship to the space they inhabit. They tend to catch the eye, so we need to think of them as focal points of our design and think carefully about their placement. They should not be competing for attention with lots of other key features; a modestly sized garden tree is going to get lost mixed in with large rhododendrons, squashed up tightly to a summer house. If it is planted close to another key feature of the garden, it must complement it. A perfectly chosen and placed tree overhanging a still pond is pleasing because of the very contrasting forms of the two features and the way the tree is reflected in the water. Two or more trees might help frame a view. Or a tree could stand more on its own. Trees have a great capability to own their spaces when their size and form correspond well to the space they have around them. Such a tree is sometimes described as having *domain*.

This tree's size, shape, colour and position give it *domain* in the space surrounding it.

Tall shrubs

Shrubs that are of head height or higher are going to screen undesirable views, such as the ugly garage next door, and provide a physical barrier to wind, perhaps noise to a certain extent, and to movement about the garden. They can also have a strong effect on how it feels to be in a certain part of the garden. A narrow path edged with tall shrubs is likely to feel claustrophobic and uncomfortable. The same shrubs might work well with a much wider path. We tend to feel very comfortable sat in a spot that is surrounded on three sides by taller shrubs but open to a view on the fourth side. The geographer Jay Appleton attributed this effect to our evolution of a need to feel safe and protected but still able to scan the world around us for possible hazards. This 'prospect-refuge' idea has probably always been part of good landscape or architectural design but it is interesting to think about why such arrangements are pleasing to us.

Medium shrubs

Shrubs of about 3–5 feet provide a physical barrier to our movement but no visual barrier unless we are sitting down. They might be appropriate for emphasising a path, subtly directing attention in a certain direction, or organising the garden into sections.

Low planting

The effect of any planting of less than three feet is going to be in terms of colour, texture and structure rather than having a visually or practically organising effect on the garden. It can also support and complement the larger planting, the classic underplanting that gardeners have long used. How to successfully use planting in this way will be discussed in the section on detailed planting plans.

The planting either side of this beautiful winding path, in a garden designed by Jennifer Gay, is the perfect size for the width of the path. Higher planting would create a claustrophobic effect.

plans. Some ideas might need to be adjusted a bit once you have researched what is available and feasible for the conditions you are dealing with, but it is much easier to approach the details when you have a clear idea of the effects that you are aiming for.

Detailed planting plans

Now you can get into the detail of the exact plants you are going to use, their positions and their quantities. Pick a defined area from your schematic to start with and work through the following steps. The ideal scale to draw planting plans is usually 1:50. A very detailed and precise perennial planting might be better shown at 1:20. For a smaller garden, 1:100 might be sufficient and allow you to get the whole planting design on one sheet of A3. Otherwise, it is fine to split the planting plan with separate drawings for different areas. Copy the shape of the planting area over onto a new sheet and resize to your chosen scale (a uniform 200 per cent if you are going from 1:100 to 1:50). In this case, it might be a good idea to produce a version of the outline design with the planting areas identified and labelled.

2. Functional planting

In the previous stage, you should have marked the rough position of any functional planting, such as a windbreak or some visual screening, that is required. If you are working on areas of planting that include a functional requirement, start with that. Choose the plant or combination of plants that is going to fulfil the functional requirement. Chapter 5 should help with ideas for plants that are particularly useful in coastal gardens or refer to the section on windbreaks and shelter belts earlier in this chapter. Once you have chosen the plants you want, draw them on to the planting plan, at your chosen scale and at their mature size. You could draw shrubs at about two thirds of their mature size if you want a quicker effect, accepting that it might start to look a bit overplanted in the later years of the planting, and might require a bit of thinning out. Or you could just plan to plant larger mature specimens but be aware that larger specimens are often quite a lot more expensive and do not always establish as quickly.

3. Focal points and trees

You have marked on your schematic planting design some points for focal point planting. These are specimen plants or planting combinations that are particularly eye catching because of their size, colour, form, or a combination of these factors. It is likely that most of the trees you want to plant will fulfil this function, unless they are part of a shelter belt or orchard. If one of these focal point plantings falls in the area you are working on, now is the time to choose what it will be and place it on the design. You could start by looking through Chapter 5 and seeing what catches your eye. Then perhaps an internet search could give you some ideas for plant combinations that would work with the plant you have chosen, to stand out and set the mood for that part of the garden. Check that it will work in the spot you have chosen. Will it get enough sun? Is it able to deal with the exposure to coastal conditions in the position you want to put it? Does it work with the general style of planting you will be using in this area? Once you have decided on the planting and the position, draw it, to scale, on your plan.

This combination of *Agave americana*, *Chamaerops humilis*, and *Cordyline australis* is a focal point in this very exposed seafront garden.

4. Structural planting

Now you should think about the shrubs that will form the framework for the area you are designing. These are likely to occupy the medium range of heights, smaller than the tree, larger than the perennials, roughly somewhere between knee and shoulder height. These forms are going to stand out in the design, and you need to think carefully about how to arrange them harmoniously. They might form the backdrop to the rest of your planting. This is quite a traditional way of organising a bed, with the large shrubs at the back and the small shrubs and perennials at the front. In this case you should think about the foliage and flower colour. Strong, hot colours, like bright red foliage, are going to seem to come forward, looking closer than they are and shortening the perceived depth of the bed. They will also attract more of the viewers' attention, which might be a very useful effect if you are trying to distract from some ugly nearby building or view. Or if they have a more standard green foliage or pastel flowers in cool colours, the effect would be more receding, acting as a background to the planting in front of it and increasing the perceived depth of the bed.

COLOUR

When we talk about plants that go well with each other, we are very often talking about a pleasing combination of colours. But what colours go well together? Most people have some opinion on, at least, which individual colours and colour combinations they like, even if it is just a matter of knowing it when they see it. But when we are required to pick colours as part of a design scheme, anxiety about making the wrong choice can creep in and interfere with our natural sense of what we like. It is a bit like being faced with a complex restaurant wine list when we are keen to impress our companions. We forget about what we like and instead become preoccupied with the fear of looking ignorant or lacking in taste. But our taste in colour, like our taste in wine, or food, or music, is very personal and influenced very much by our personal history of previous encounters with colours and our associations with them. Other parts of the design process are more about practicalities, and it is a good idea to follow a carefully considered process, but when it comes to colour you are freer to please yourself. So our first advice on using colour well in your planting design is to forget the rules, forget received ideas of good taste, and just go for colours that excite you, that bring you joy, that make you feel peaceful, or excited, or just glad to be alive, because that is what gardens and flowers should be about and only you know what colours do that for you.

However, there is such a thing as colour theory, and it can be helpful to understand it. Often the hardest part of the design process is making the first few decisions on the blank design and having got the first plant in place, colour theory might help you think about what colour you might be looking for in the next plant.

If you are producing the design as a professional garden designer, your own feelings about colour are less important. Later in your career clients may be coming to you for your own style, but in general, as designers our first duty should be to interpret and make the best of our client's own taste. We can ask them about what colours they like but then they will usually be looking to you for tasteful and pleasing schemes based around these colour preferences. This is where a little colour theory can help. Just keep in mind that it is not an exact science and good design often bends or breaks the generally received rules about good use of colour. This is particularly true in planting design as our colour combinations almost always take place against a backdrop of foliage colour, usually some shade of green. This can allow us to get away with some colour combinations that would be considered clashing if we were using them in interior or fashion design. Good examples of these successful clashes can be found in traditional cottage gardens; here the often riotous combination of theoretically clashing colours is what many people first think of when they imagine an ideal garden.

The cool colours of the lavender and pastel yellow rose recede and create more sense of depth along this path.

Alternatively, you might want to mix the structural shrub planting into the rest of the planting, creating planting masses in different parts of the bed, some forward, some further back. This can be more interesting than the traditional approach, but care must be taken to maintain a sense of balance. It works well if, for instance, two or three larger shrubs on one side of the bed are balanced by four or five smaller shrubs on the other side, as if they are balanced on an imaginary see-saw. The fulcrum of this balance need not be the centre of the bed; perhaps they balance around a focal point tree planting, or they form a balanced group in one corner of the bed with drifts of perennials dominating the rest of the bed. The point is to consider where the masses of the planting are and to arrange them in a harmonious way. It is also often a good idea to surround the highest shrubs with some slightly smaller shrubs before transitioning to the perennials and ground cover. Many shrubs don't look right sticking up on their own amongst much smaller planting, but some do – it just needs to be done with care.

It is also a good idea to think about foliage types and form when grouping shrubs. A group of shrubs that have the same foliage texture and colour will have a more imposing effect than one with shrubs of varying texture and colour. And some foliage types don't work so well *en masse*. Spiky leaved plants such as phormiums and agaves don't tend to look right clumped together and are better paired with some softer forms, such as airy, flowing grasses. But you may want to group them for a special effect. None of these are hard and fast rules, just things to be careful about.

Colour theory usually starts with a colour wheel. The basic idea of this is that opposite colours are complementary. So, for example, blue and orange are considered complementary. What this means for us is that some bright orange flowers will look particularly intense and will stand out particularly strongly against a background of intensely blue flowers. This complementary effect is sometimes interpreted by people as clashing; what they are reacting to is the dazzling nature of complementary colour combinations. They are very eye catching and can be highly effective but should usually be used sparingly. Sometimes a safer bet is to choose the two colours on either side of the complementary colour. For example, blue with a red orange and a more yellow orange. This creates a more subtle complementary effect.

Another useful idea from the colour wheel is that of analogous colours: the colours either side of the colour you are trying to match with. So, the colours analogous with blue are blue-green and blue-violet. In contrast to the vibrancy of complementary colour combinations, analogous colours create a harmonious effect. There are a lot of these analogous combinations to be found in coastal environments: the intense blues of sea and sky mixed with the greens and blue greens of a lot of coastal foliage, or the reds, oranges and yellow of the sun setting over the ocean. It can be very effective to echo these colours in our planting schemes.

One final idea from the colour wheel that might be useful is that of triadic colours. These are three colours that are evenly spaced around the wheel. For example, purple, green and orange. The effect of these is related to the complementary colours and creates a vibrant effect. As with the complementary colours it is best used sparingly; overuse of this effect can be a bit jarring and chaotic.

Of course, one of our main aims is to avoid colours that clash. As we have noted previously, personal feeling comes into this, but in general, it is not that particular colours clash in all their forms, but that some shades of those colours don't work well together. In the above examples of complementary, analogous, and triadic colours we are talking about the bright, pure hues of those colours. And, in general, bright colours will work best with other bright colours. Problems can arise when you mix a muted version of the colour with a bright colour. Also colours with cool undertones (colours with blue-greens, blues and purples mixed in) can clash with colours with hot undertones (the reds, oranges and yellows). But this is less definite in gardens than it is in other design projects. The muted purple of lavender seems to go very well with the bright yellow of rudbeckia. This is because the diluting effect of foliage and the ever-changing light in a garden complicate the effects of colour a great deal. But it is useful to think about a general theme for your colours, perhaps hot and intense, or cool, calm and pastel. Or you could use one or two main, strong colours against a background of neutral browns and greys.

It is also often a good idea to have a dominant colour, making up perhaps 60 per cent of the scheme, with a good, reliably harmonious supporting colour making up perhaps another 30 per cent, with a final, more intensely contrasting colour used much more sparingly as an accent. In general, the best advice is to be aware of some colour effects we have discussed, look out for great colour combinations in gardens, online or in the gardening press, and do not be afraid to trust your instincts and experiment.

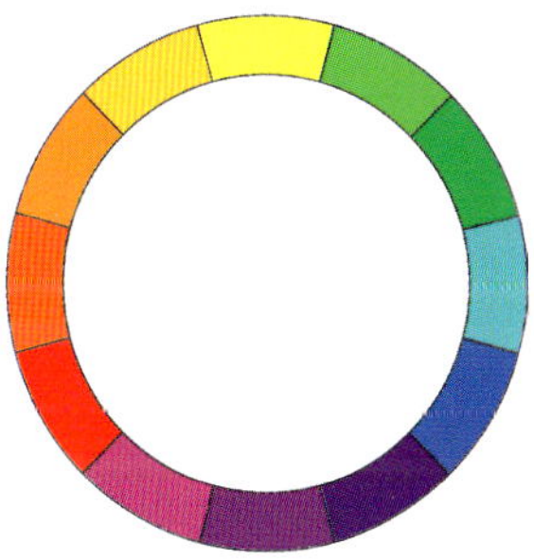

The colour wheel.

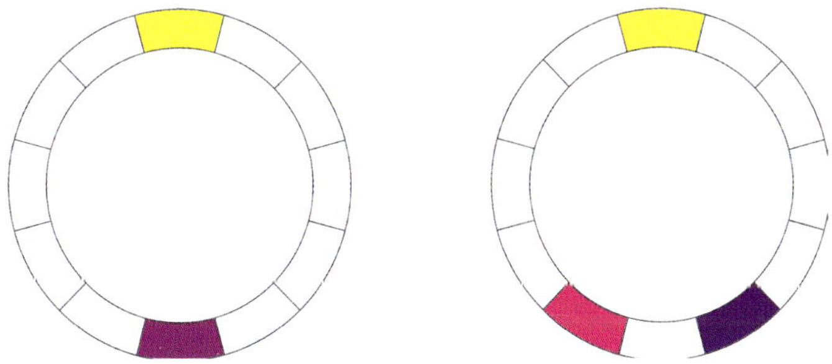

Complementary colours (left) and split complementary colours (right).

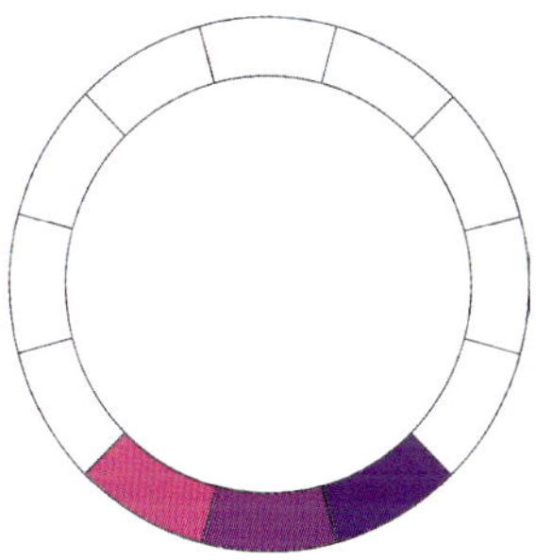

Analogous colours.

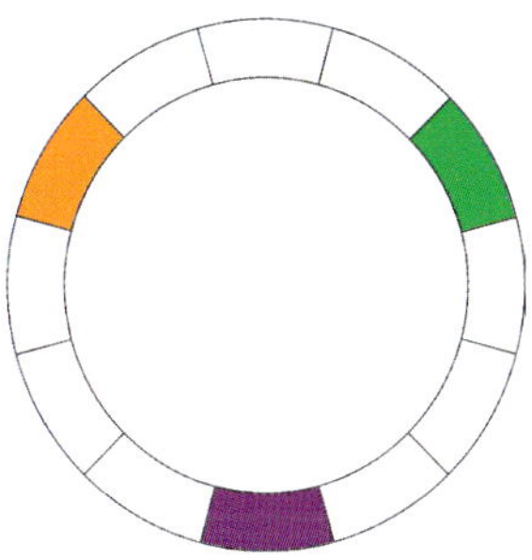

Triadic colours.

5. Perennial planting and ground cover

Now you should consider the rest of the planting. This is likely to consist of perennials and annuals that, although they may contribute an architectural effect through their form, are contributing mainly colour and texture rather than practical or structural effects. A good principle to start with here is to have fewer different plants than you might anticipate but to plant each in larger quantities. A piece of advice often given to garden design students is that they can improve their planting plans by 'halving the species, doubling the quantities'. A lot of gardens do not live up to their potential because the planting has been done in a piecemeal way and plants have not been used in sufficient quantities. A single plant is bought on impulse during a garden centre visit and then fails to have any impact on the garden other than to contribute a sense of incoherent messiness. So, plant in groups of five, seven, fifteen or even thirty or more if the size of the bed permits it. Being bold and using broad strokes is usually the way to go. We also usually, if we are dealing with smaller quantities, plant in odd numbers. The idea is that we tend to mentally split an even number into two groups, and this reduces the coherence of the design. Perhaps, but whatever the reason it does seem to work better to use odd numbers of a particular plant. It is not, however, important for quantities greater than about ten.

As you work on the planting plan you must also keep thinking about how the planting will change over time. If you are looking for a particular colour combination, are they going to be in flower at the same time? Thinking about the whole garden, will it have sufficient year-round interest? This is easier to achieve in a larger garden and it is ok for a smaller garden to be at its best at a certain time of year but there should be something of interest at all times of the year. Be wary of plants that only do one thing, such as forsythias, which produce tremendously vibrant yellow flowers in early spring but then look a little drab for the rest of the year. Consider if a plant will contribute enough to earn its place, particularly in a small garden.

You could start with one of the plants you already have on your plan, one of the focal points or structural shrubs. What would work well with it? What colour? What texture? What size? Look through our plant suggestions in Chapter 5. Or search online for planting combinations with the shrub or tree you are starting from. Once you find one you like, check that it will work in the conditions present in the part of the garden you are working on, such as the amount of exposure to coastal winds or the amount of shade; then, if you are confident that it will work, draw it to scale on the plan. Then carry on, finding the next combination and the next, fitting them together like a jigsaw. It is a painstaking process but if you like plants it should be a pleasurable and interesting one. We have included several planting schemes in this book that might give you a head start.

If you find it difficult to imagine how plants will look together you could just get an idea of the number of typically sized perennials you will need and make a list of plants that will fit your rough colour scheme and work for the conditions they will be planted in. Then, rather than drawing every individual plant on the plan you could visit a garden centre and start making a collection of plants from your list and see how they look together in the trolley. This can then be tweaked when you get home and lay them out ready to be planted. This method is perhaps best suited for use in your own garden rather than in a professional context. But sometimes even as professionals we do not draw every perennial or annual individually on our planting plans. If we are working on very large-scale planting or on a meadow, we might just indicate the area, the plants we want to include and what percentage each plant accounts for in the mix.

MAINTENANCE

Our choice of plants is going to have a drastic effect on the amount of work the garden will require to keep it looking good. In professional practice you will often be asked to produce low maintenance gardens as your clients are quite likely not to be keen gardeners.

The maintenance input is inevitably higher in the first few years after planting as the plants fill in the spaces allowed for them. The beds will require more weeding during this time than when the planting is mature. You can reduce this requirement by planting larger specimens, placing greater emphasis on shrubs than perennials, and using shrubs that require less pruning. You could also use a mulch which helps to suppress the growth of annual weeds. We have included advice about mulching and suggestions for shrubs that require minimal pruning in the techniques section of this chapter.

A worked example

DEVELOPING THE PLANTING PLAN

As we did with the outline plan in Chapter 2, we will now show how we produce a planting plan using a step-by-step process. A planting scheme can seem to be rather dauntingly complex with functional and horticultural considerations needing to work with your ideas of texture, colour, form, and focus. Breaking the task down into a logical sequence of small steps makes the task much more approachable.

For some smaller areas of planting, you may not need to use every step in this process. You may only need to draw the area to scale to work out the quantities and a few attractive plant combinations. But for any larger, more complex areas this process will give you the best chance of creating an attractive planting scheme. If you find yourself struggling for a good combination with any plant, an internet search for something like 'planting combinations using cistus' should give some inspiration.

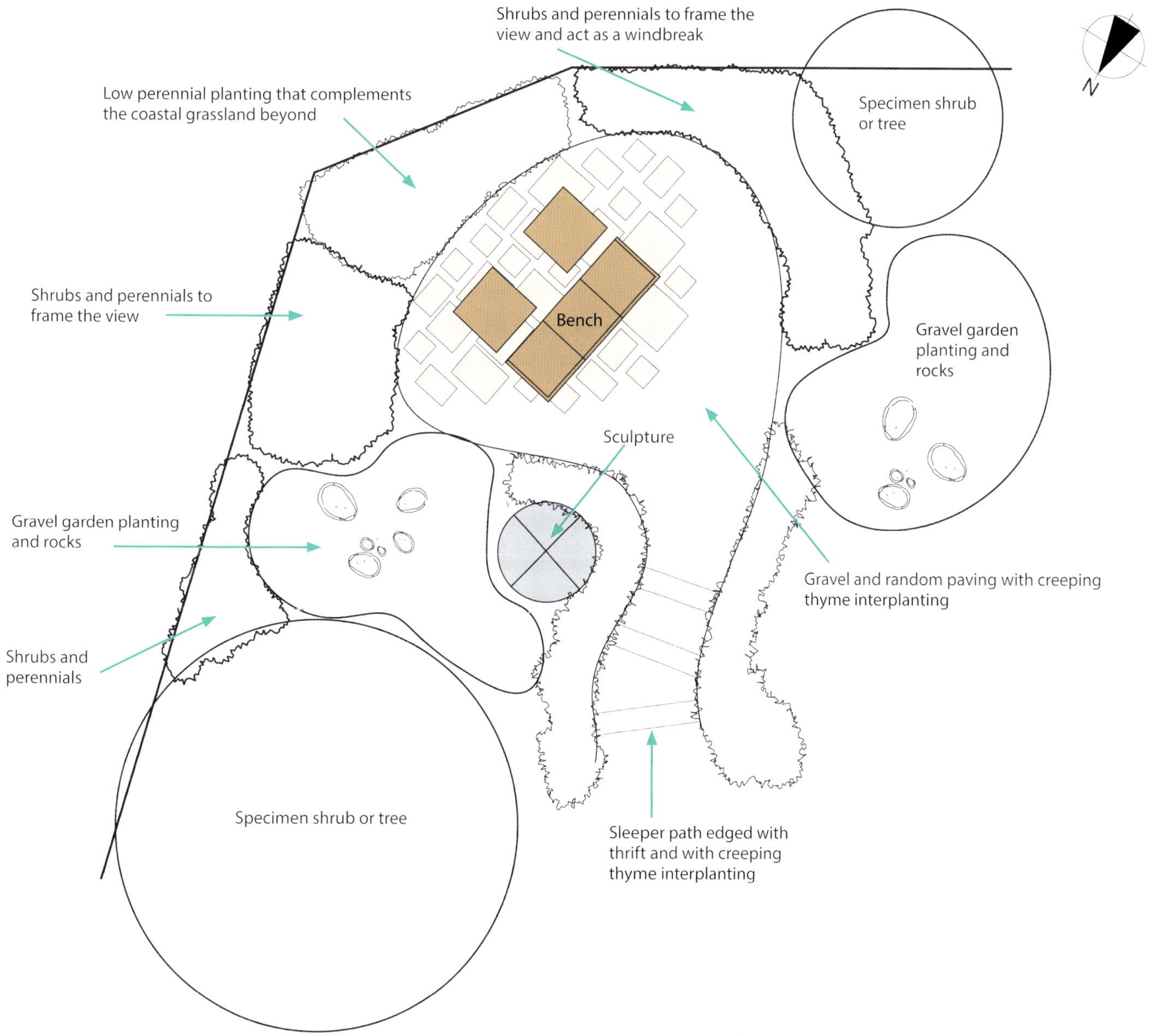

Step 1: Schematic

We sketch in the focal point plantings, the functional plantings, and the planting styles for the area we are working on.

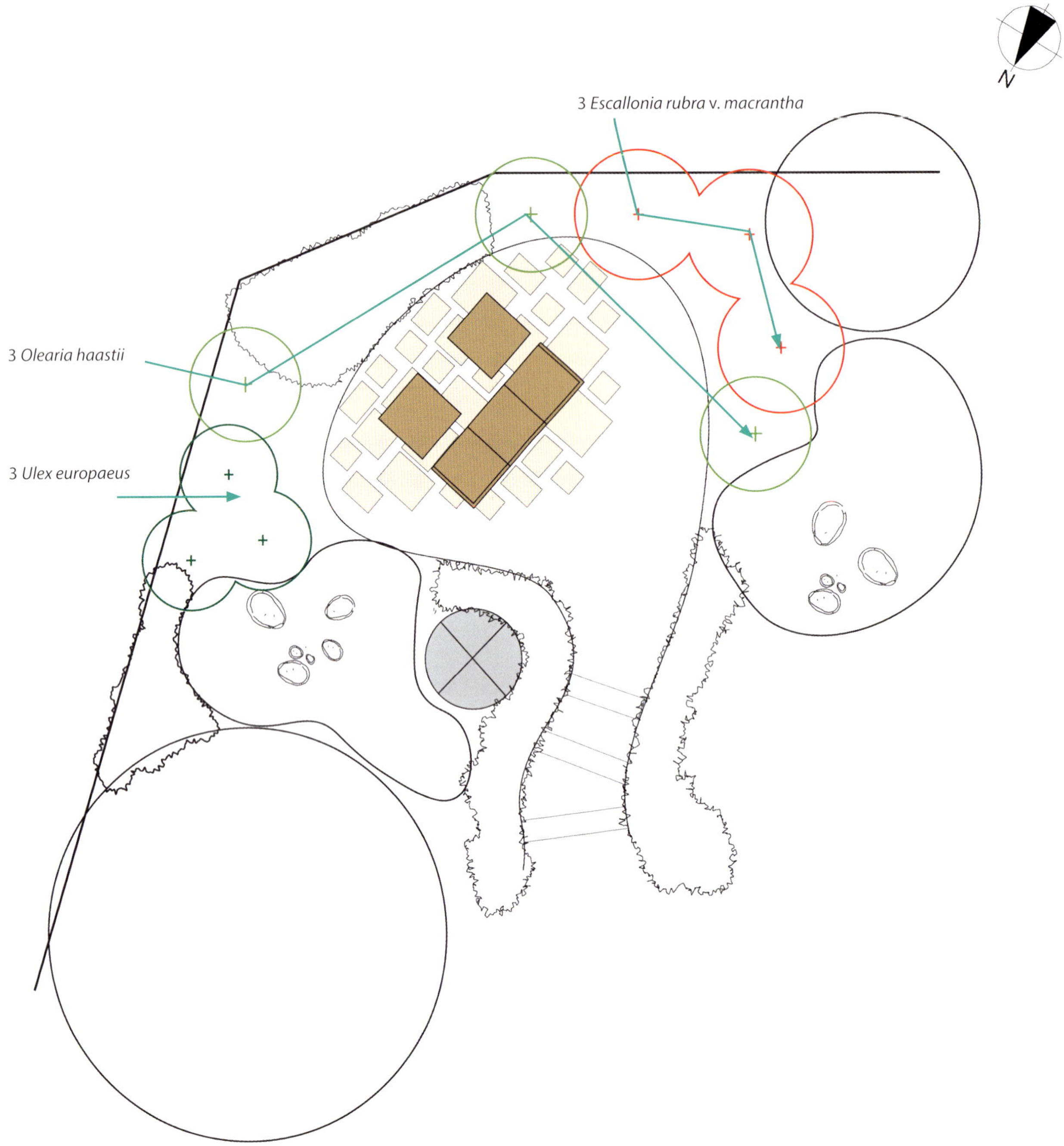

Step 2: Functional planting

We place things such as windbreaks or visual screens and decide on the plants we are going to use.

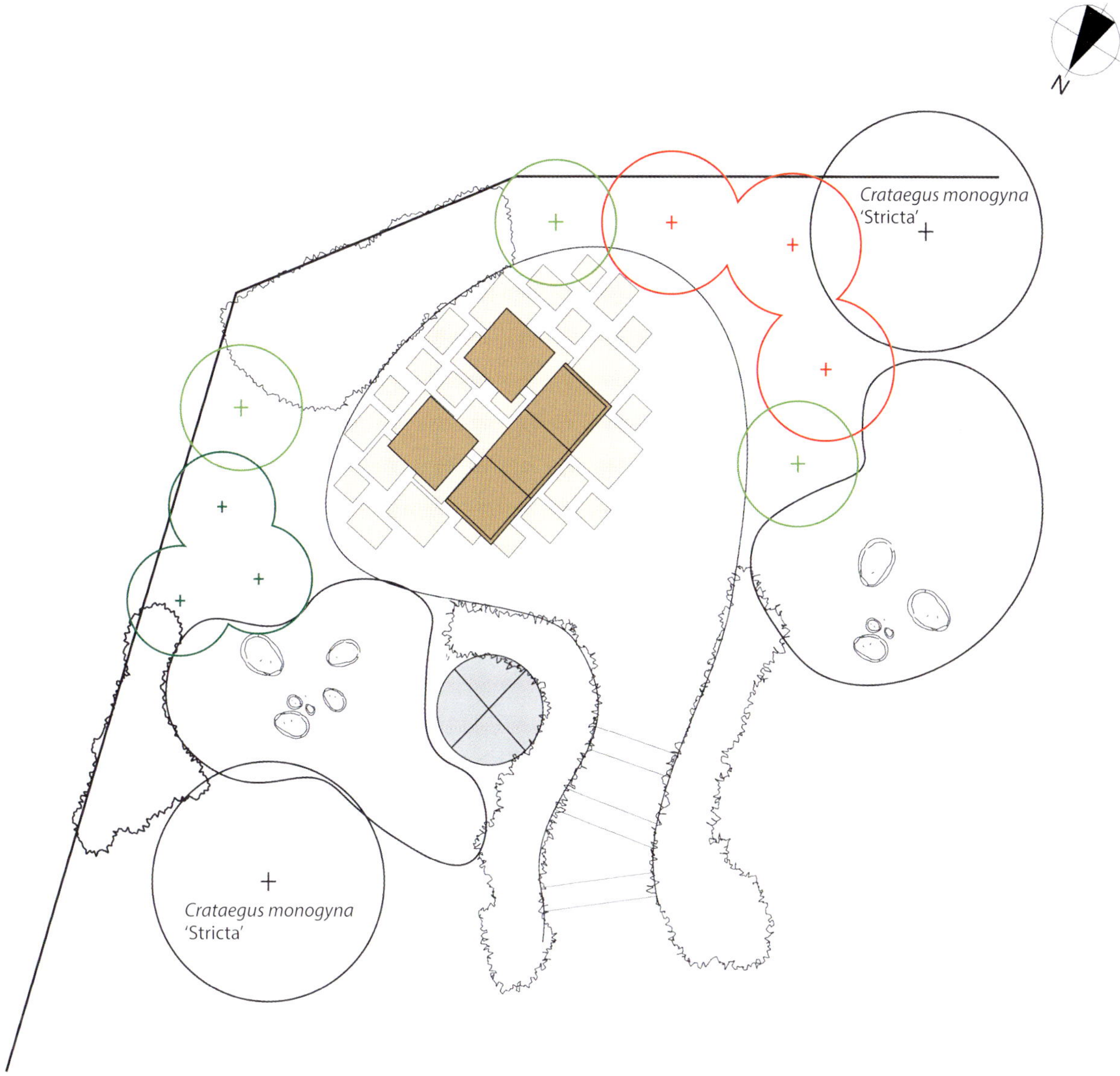

Step 3: Focal points and trees

We decide on the position and types of focal point plantings and other trees.

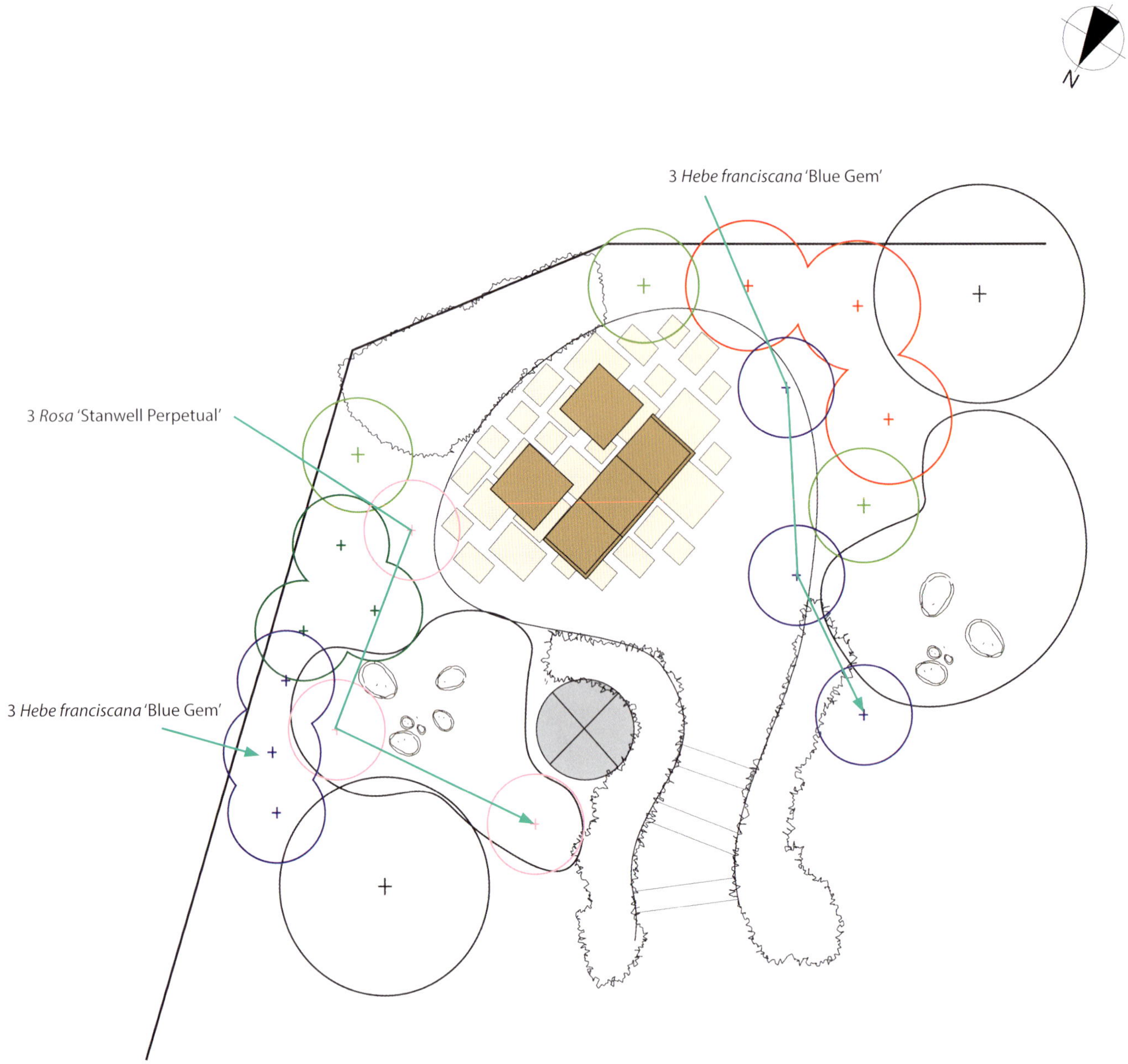

Step 4: Structural planting

We place the backbone shrubs, thinking about the three-dimensional effect of size and texture.

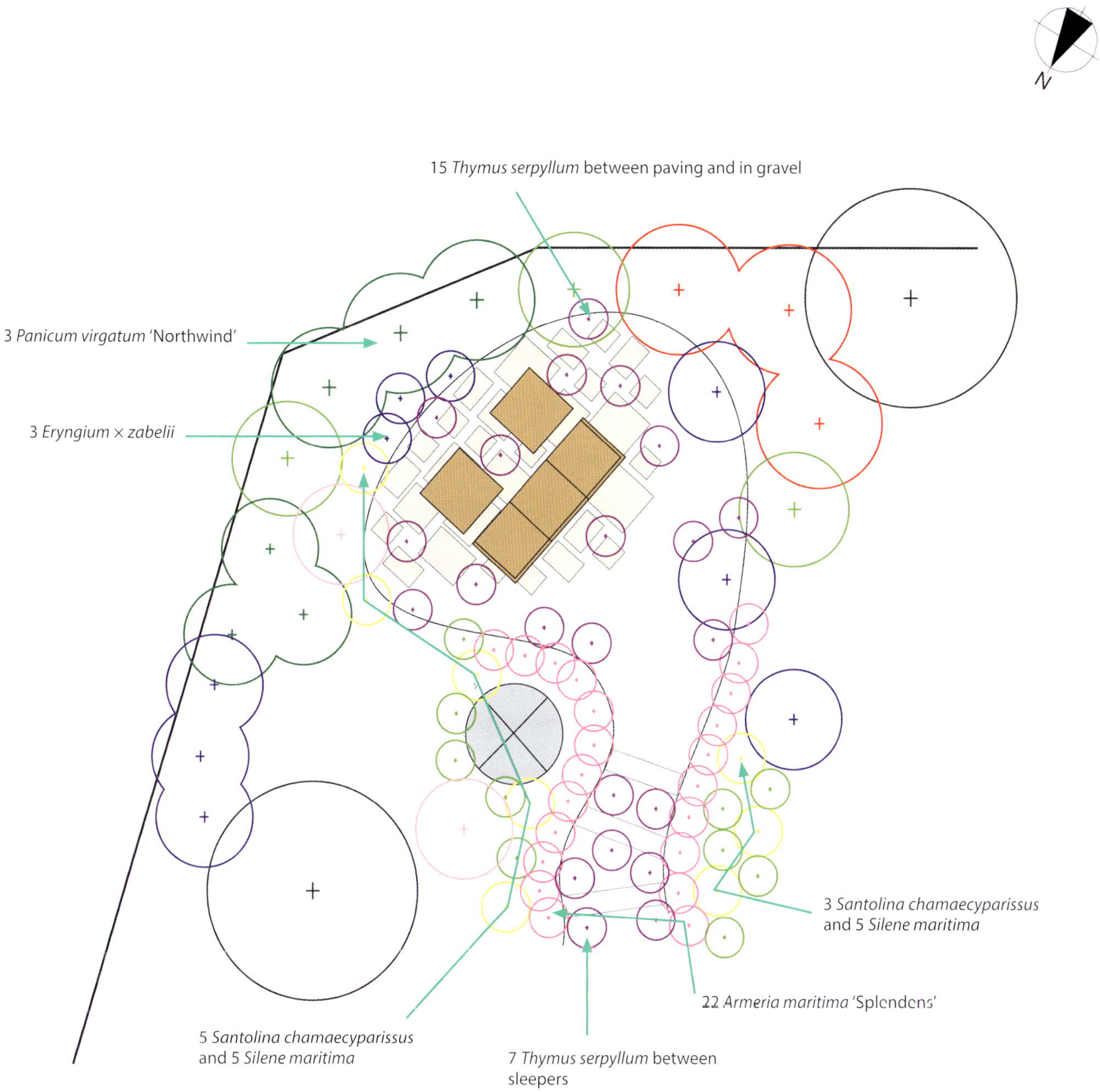

Step 5: Perennial planting and ground cover

Thinking about good combinations of colour and texture, we fill the rest of the space with perennials, annuals, and ground cover.

THE PRODUCTIVE COASTAL GARDEN

The coastal climate can be very hard on both vegetables and fruit. Success in the cultivation of these requires a little planning and effort. Sandy soil requires lots of organic matter to be well dug in. This aerates the soil, adds nutrients, and increases moisture retention. Plants being scorched by both wind and salt can be a major problem. Research has shown that beet, kale, spinach and asparagus have a high tolerance of salt and that broccoli, cabbage and tomatoes have a medium tolerance. This should be borne in mind when deciding where to grow vegetables. The use of cloches or a plastic tunnel will add shelter from the wind. An alternative is to surround the vegetable garden with shrubs that can cope with the coastal conditions. Do bear in mind that most vegetables prefer plenty of sunshine. And, because the wind is likely to cause problems to tall varieties of plants it may be wise to choose low-growing cultivars to minimise the problem.

Fertilisers

As sandy soils are poor at retaining nutrients it is important to feed the plants on a regular basis, especially during rainy periods when the nutrients are washed away. An inorganic, balanced fertiliser can be used at the start of the growing season as it is fast acting, but it is lost from the soil quite quickly, especially in wet weather. A balanced organic fertiliser is an alternative, but it does not produce an immediate response. Crops such as tomatoes should be fed a soluble, high potash fertiliser once the first tomatoes start to form.

Watering

Watering also tends to be particularly important in coastal gardens as sandy soils are quick to dry out. Plants should be given a thorough soaking around the roots when there are signs of dryness. Do not use small quantities daily as this will encourage surface rooting which exacerbates the problem. Ideally, water between six and nine o'clock in the morning. This will give time for the plants to absorb water before the sun becomes too hot. An alternative is to water in the evening, but this may encourage slugs. After a day of rain apply a 2-inch (5cm) layer of mulch around the crops, but not up against the stems. This will require renewing at intervals as the mulch is taken into the soil by worms.

Herbs

Dry soils with a low nutrient level are the natural habitat for many herbs and these are a good choice for the productive coastal garden. Do bear in mind, however, that the majority of herbs prefer a very sunny spot. Garlic, chives, sage, oregano, thyme, lavender and rosemary should all do well. Mint will be happier in a shady, more sheltered spot.

Raised beds and containers

On a smaller scale it is perfectly possible to grow good vegetables in raised beds or containers. It has advantages, particularly in a coastal garden, because the gardener has a greater control of the soil makeup than they have in the open garden. Containers and

A beautiful and productive vegetable garden on the Isle of Mull.

The tomatoes are bush cultivars rather than cordon cultivars which are grown on a single stem and have all side shoots removed. Bush cultivars will not get too tall, and this will reduce any problems with the wind. No staking is required, and the tomatoes have been planted in a large container. The cultivar is 'Losetto' which has good resistance to blight and a long cropping season.

raised beds do dry out more easily but using both garden and commercial compost will improve the water and nutrient retention. Larger containers dry out less quickly than smaller ones but are heavy, and an advantage of growing in containers is that they can be moved to a more sheltered spot if bad weather is forecast. But they are also easily covered with garden fleece or other suitable material on very windy days. If the decision is made to grow the vegetables in containers or raised beds, choose crops that do not take up a lot of space and are less readily available in the supermarket.

The medium in the containers consists of a large amount of extremely well-rotted garden compost and ordinary garden soil. They are watered regularly and fed with a high potash feed once a week. The result is some healthy leeks and parsnips. The leek cultivar is 'Sultan'. It has good resistance to rust and will be ready for harvesting from November. The parsnip cultivar is 'Tender and True' which has very good flavour and good resistance to canker. Note the *Griselinia* hedge, which is acting as a windbreak.

The onions were purchased as sets and planted in a raised bed and smaller containers. They were fed with a high potash feed on a regular basis and have reached the stage when they should be lifted and left to ripen. When the foliage is dry and papery the bulbs will be stored in a cool dry shed. The cultivar is 'Hercules' which has good resistance to bolting and will be ready for harvesting from late August.

Growing fruit

The approach to growing fruit is similar. Strawberries are low growing, easily sheltered from the wind and perfect for raised beds. On the other hand, raspberries like a lot of moisture and this is not typical of coastal soils, so they are not so easy to succeed with. Goji berries can cope in windy coastal areas and the morello cherry will grow in the shade and it should be possible to grow it as a bush, or train it on a sheltered wall. Goji berries are best trained on wires along a fence (be careful of the spines) and should be fed with a high potash feed every two weeks once they are growing well. Blackberries can also be wall trained in a shady position and produce successful crops.

Fruit trees are often unsuccessful in exposed coastal gardens. Good staking is essential, and the effect of strong winds can be reduced by planting on the leeward side of a suitable windbreak. Strong winds reduce visits by pollinators and for later crops autumn storms can cause a lot of fruit drop. The trees should be planted in as sunny a position as is possible and on dwarf or semi-dwarf rootstocks. Buy from local growers, if you can, as the trees will be more used to the conditions in your garden. Ideally, the trees should be planted in late spring as this gives them the maximum time in which to get established. It is important to keep them well watered until they are fully established. This could be for two years.

The blackberries are a thornless variety, 'Oregon Thornless'. Once the fruit has been harvested the fruiting stems should be cut to the base and the new ones tied in. Any long side shoots can be reduced in length.

PLANTING TECHNIQUES

Soil preparation

'Right plant, right place' is a well-known saying in horticultural circles. This topic is looked at in detail in Chapter 5. 'The answer lies in the soil', another well-known adage, requires some explanation. If a couple of handfuls of garden soil are taken from a few inches down, added to a coffee jar of water and shaken well it will settle out into layers. The bottom layer will be grit and sand, loam will constitute the second layer and clay will form a top layer. These are the ingredients of soil and what we are looking for is a good mixture of the three to produce a well-drained, moisture retentive soil. This is the ideal and if this is what you have then you are incredibly lucky. Look after it and add plenty of garden compost on an annual basis to maintain its organic content.

Clay soils

If your soil has a lot of clay, then it has good water retention and is very fertile as the particles of clay hold onto important minerals. However, it can be difficult to work and is slow to warm up in the spring. Digging the soil at the start of winter and leaving it in lumps for the frost to break down does help to make it more workable, as does the addition of quantities of compost and horticultural grit. Do not walk on wet clay soil as this leads to compaction which is the opposite of what you want.

Sandy, gritty soils

If your soil has a lot of sand and grit, then it drains quickly, and water-soluble minerals are poorly retained. However, it warms up quickly and is easy to work even when wet. Addition of quantities of compost will increase the water retention.

Soil pH

It is important to know the pH of your soil; this is easily determined by purchasing a pH measuring kit from the local garden centre and following the simple instructions. A pH of seven is neutral, less than seven is acidic and greater than seven alkaline.

Hydrangea macrophylla cultivars can cope with coastal conditions in both acidic and alkaline soils. This cultivar is *H.* 'Immaculata'. Its height and spread are 1.5m (4.5 feet).

TABLE 4.1 SUGGESTED SUITABLE PLANTS FOR COASTAL SITUATIONS BY SOIL TYPE

Alkaline soil	Acidic soil	Heavy clay soil	Light sandy soil
Berberis cultivars	*Astelia chathamica*	*Berberis* cultivars	*Buddleja globosa*
Brachyglottis 'Sunshine'	*Berberis* cultivars	*Crataegus laevigata*	*Brachyglottis* 'Sunshine'
Buddleja globosa	*Buddleja globosa*	*Kniphofia* cultivars	*Cistus* cultivars
Ceanothus thyrsiflorus	*Ceanothus* cultivars	*Pinus mugo*	*Convolvulus cneorum*
Cistus cultivars	*Hydrangea macrophylla*	*Rosa rugosa* cultivars	*Cortaderia* cultivars
Crataegus monogyna	*Phormium tenax*	*Sorbus aucuparia*	*Elaeagnus* cultivars
Elaeagnus angustifolia	*Pinus contorta*	*Vinca minor* cultivars	*Eryngium* cultivars
Escallonia cultivars	*Pinus mugo*		*Genista hispanica*
Euonymus japonicus			*Hippophae rhamnoides*
Fuchsia magellanica			*Lavandula* cultivars
Hebe × *franciscana*			*Phlomis fruticosa*
Hydrangea macrophylla			*Phormium tenax*
Santolina chamaecyparissus			*Pinus contorta*
Spartium junceum			*Rosmarinus officinalis*
			Spartium junceum
			Tamarix tetrandra
			Ulex europaeus
			Yucca cultivars

This information is a very important guide in the choice of which plants you are going to grow. Plants suitable for coastal gardens are discussed in detail in Chapter 5 but as a guideline some recommendations, sorted by preferred soil types, are included here. Some of the suggested plants are suitable for more than one situation. Refer to Chapter 5 for further information such as whether the plants prefer sun or shade.

Ground preparation

'Spend as much time on the hole' is yet another well-known horticultural saying and is telling us to take our time over the preparation of the planting hole. Container-grown plants can be planted at any time of the year if conditions are right. However, in midsummer it can be too hot and the soil too dry and hence difficult for the plant to settle in quickly. Similarly in winter soils may be too wet or frozen and such conditions are not suitable for planting.

When preparing planting areas, remove any unwanted turf. Use a half moon to cut the turf into manageable sizes and use a sharp spade to skim off the turf as thin as possible. An alternative is to hire a turf stripper from a tool hire shop. The turf that you have removed should be stacked upside down and left to rot. This produces high quality loam which can be used in plant containers or just added to the planting areas. The soil can then be dug and any perennial weeds such as dandelions, convolvulus and ground elder removed. Weeds such as these will regrow from any small piece of root left behind so take your time and remove as much as is possible. If the digging is done on a sunny day, then annual weeds can be left to lie on the surface to dry in the sun and wind. If time permits, leave the ground alone for a couple of weeks to allow germination of annual weeds and these can then be hoed off and left in the sun. Annual weeds can be put on the compost heap but not the perennial weeds.

When digging the planting hole, ensure that it has a diameter about 30cm (1 foot) wider than the container and deep enough for the top of the root ball of the plant to be at existing soil level. If the soil at the bottom of the hole is very hard, loosen it with a garden fork. Ideally the plant should be well watered whilst in its container the day before planting or at least one hour before. If your soil is poor, add well-rotted compost or soil improver to the excavated soil and mix well. Add some compost or soil improver to the base of the hole and, if the soil is not very moist, fill the planting hole with water and allow to drain. Give a light dressing of a general fertiliser such as fish, blood and bone to the base of the hole and the excavated soil. You are now ready to plant!

Planting

Tap the plant out of its container and loosen some of the roots if it is pot bound. Place the plant in the hole so that the top of the root ball is at soil level or slightly below. Back fill the hole with the excavated soil and firm with the hands. Top up to the correct level and tread down gently. Water well.

Bare-rooted plants are a cheaper option than container grown plants. They are dug up in the nursery when dormant and it is important that the roots do not dry out. If you take possession of them before you are ready to plant them, they should be heeled in, that is, temporarily place them in the trench and cover the roots with soil. When you are ready to plant them in their permanent spot, prepare the planting hole as above and check that the soil level and the soil mark on the stem of the plant will coincide when planted. A piece of wood laid across the hole is useful in ensuring that this is the case. Any damaged roots or very long ones can be cut back to about 30cm (1 foot). Ensure that the roots are well spread out but do not force the roots into the planting hole or bend them back to make them fit. Firm in with the addition of compost or other improved soil and water well.

When planting, especially trees and shrubs, it is sensible to space them a good distance apart. A rule of thumb is to add the mature heights of the two neighbouring plants together and divide the total by two for trees and by three for shrubs. If the mature heights are not known, depending on the size of the plant, trees should be planted 5–8m apart (16.5–26 feet), shrubs 0.5–1.5m (1.5–4.5 feet) apart and perennials at a rate of 3 to 5 per square metre (11 square feet).

Tree and shrub staking

Staking of trees is important, especially in exposed situations. For bare-rooted trees a stake should be inserted to a depth of about 60cm (2 feet) about 10cm (4 inches) from the trunk. Ties can then be put in place at the top and bottom of the stake. Large shrubs may also require a stake. In this case tie the stake to the main stem of the shrub. For container-grown trees the stake should be placed at an angle of 45 degrees and inserted in the soil so that it avoids damaging the root ball. The tie can be attached halfway up the trunk of the tree. For large trees or on very windy sites two vertical stakes can be used attached to a cross bar to which the tree is tied. The two stakes should be placed so as not to damage the root ball. In all cases the ties should be inspected at intervals to check that they are not cutting into the bark of the plant; if it is too tight, loosen it.

A newly planted *Crataegus monogyna* 'Stricta' suitably staked for a windy site. The staking should remain in place for two to three years.

PLANT MAINTENANCE

Mulching

Mulch, an organic layer added to the surface of the soil, will enhance the appearance of the planted areas in the garden. Ideally it should be done in late spring when the soil is moist and warm. The layer should be a minimum of 5cm (2 inches) thick. It is far better to mulch half of the garden with a 5cm layer than the entire garden with a 2.5cm layer. The other half can be mulched the following year. Mulch should be kept away from the trunks of trees and stems of shrubs and perennials. It can consist of home-made garden compost, well-rotted manure (although this may contain a lot of weed seeds), or old growing compost from containers and grow bags, leaf mould and similar organic material. Personally, we do not recommend the use of cut grass as it will reduce nitrogen levels in the soil as it decays. The mulch helps retain moisture in the soil, improves the structure of the soil and suppresses the growth of annual weeds. There is no need to hoe it as the worms will take it into the soil. Mulching is excellent for the overall quality of the soil, but it is not a fertiliser, and the plants may well require feeding as well.

Feeding

Application of a suitable fertiliser will increase the quantity of plant nutrients in the soil. The three main nutrients for plant growth are nitrogen, phosphorus and potassium (NPK). Nitrogen encourages good leaf growth, phosphorus encourages healthy root and shoot growth, and potassium assists in the production of healthy fruit and flowers. There are two main groups of fertiliser: quick acting and slow acting. Quick acting fertilisers are made by the chemical industry and the nutrients are in a form that plants can use almost immediately. As the fertilisers are water soluble, they are lost from the soil relatively quickly. Slow acting fertilisers are usually organic, such as hoof and horn meal, bone meal, and blood, fish, and bone. Bacteria within the soil break them down to a form that the plants can absorb and this is a slower process. However, as they are not particularly soluble in water they are retained by the soil for a good length of time.

Pruning

Having encouraged the plants to grow well, it will be necessary to carry out annual pruning on many shrubs and some trees. The purpose of the pruning is to remove dead wood, damaged stems, and any crossing stems that are rubbing against each other; to control the quality and quantity of flower production; and to control the shape of the shrub. It is important to have a set of sharp tools to carry out this work. A good pair of secateurs is particularly important; do not waste money on a cheap pair. In addition, a pruning saw and a pair of loppers will be required, and for work on trees and large shrubs, a cord-operated tree lopper is useful. Here are some guidelines for shrubs and a few trees that are often used in coastal planting schemes.

- *Artemisia* 'Powis Castle' – does not respond well to hard pruning but does require an annual trim. Cut back to where new growth buds are visible. This should be carried out at the end of the summer or at the start of spring.
- *Atriplex halimus* – requires minimal pruning if grown as a free-standing shrub. Just a light tidy-up at the end of winter is usually sufficient.
- *Berberis* cultivars – little or no routine pruning is required. They will take hard pruning if they have outgrown their allotted space.
- *Brachyglottis* 'Sunshine' – prune after flowering by cutting back stems that have flowered to a point where new growth is starting to appear. Some gardeners prefer not to let the shrub flower and prune the stems back before the buds open. This produces an attractive mound of silvery foliage.
- *Buddleja globosa* – if it is overgrown it can be cut back to about 1m (3 feet) high. This can be done in late February to mid-March, but this will stop it flowering in that year. Ideally it requires little pruning; if some light pruning is required then it should be done immediately after flowering. The more common *Buddleja davidii* should be hard pruned annually. This should be done in late February but if done

as late as May, it will flower later than usual which will be of benefit to butterflies such as red admiral and small tortoiseshell.

- *Carpinus betulus* – requires minimal pruning. If necessary, it can have a light trim in late summer or early autumn to maintain its appearance. It is prone to bleeding if pruned earlier in the year.
- *Ceanothus* cultivars – an annual pruning is important as they do not respond well to hard pruning. Prune lightly after flowering. It is usually safe to cut back to where a new shoot is appearing.
- *Chamaerops humilis* – no pruning is required.
- *Cistus* cultivars – many do not respond well to pruning. If pruning is essential, then do so after flowering and cut back to where a vigorous shoot is growing. If the shrub becomes old and leggy, replace it.
- *Convolvulus cneorum* – will require a late prune in early spring to remove any foliage that has been damaged by the winter weather.
- *Cordyline australis* – is not usually pruned but will require the removal of dead leaves which are prominent at the base of the foliage. A sharp downward tug will usually achieve their removal.
- *Crataegus monogyna* – very long branches can be cut back in late winter if it requires reshaping. As a hedgerow plant it is very tolerant of hard pruning.
- *Elaeagnus* cultivars – prune as required in spring but do not cut into very old wood. Remove all stems that have reverted, i.e. are all green, on the variegated cultivars. Hedges can be trimmed a couple of times a year. *Elaeagnus* 'Quicksilver' can be pruned after flowering.
- *Escallonia* cultivars – prune straight after flowering, removing the flowered stems. Overgrown specimens can be hard pruned as they regenerate well from old wood.
- *Euonymus* cultivars – evergreen cultivars can be pruned hard in spring. Remove any reverted stems when they are seen.
- *Fuchsia magellanica* – prune in early spring. It can be cut back to the base or given a lighter pruning if overall height is important.
- *Genista hispanica* – requires minimal pruning. If necessary, can take a light trim after flowering but it will not regenerate from old wood. If the specimen is overgrown and untidy then replace it. The same treatment is the case for *Genista lydia*.
- *Griselinia littoralis* – requires an annual cut to keep the hedge neat and tidy. This can be done in spring/early summer or late August/September. The later date is more likely to avoid disturbing nesting birds.
- *Hebe* cultivars – pruning is not usually essential, but any winter damage can be removed in spring. Cut damaged stems back to new shoots. Dead heading, i.e. removal of old flower heads, will prolong flowering and help to maintain a good shape of shrub.
- *Hippophae rhamnoides* – requires minimal pruning. During the winter it can be trimmed to maintain a decent shape and to remove any dead wood.
- *Hydrangea macrophylla* – some of the old woody stems can be removed by cutting to ground level in spring. Remove any old flower heads by cutting back to the first pair of good shoots below them. Otherwise, little or no pruning is required.
- *Lavandula angustifolia* – cut back lightly after flowering but do not cut into old wood. Woody, overgrown specimens are best replaced.
- *Lupinus arboreus* – no pruning is required except for any bits of stem damaged by the winter. Leave a few dead flowers on the plant so that some self-sown seedlings will be produced, as the plant is not particularly long lived.
- *Olearia* cultivars – if grown as hedging will require an annual trim when mature to keep tidy. This can be done in late summer. After flowering, the top of the hedge should be narrower than the base to avoid the bottom of the hedge becoming bare. *Olearia* × *scilloniensis* is usually grown as a free-standing shrub and should be given a light pruning after flowering. Cut back to new growth to produce a bushy plant.
- *Phillyrea angustifolia* – requires minimal pruning. Light trimming can be carried out after flowering to remove any shoots that are spoiling the appearance of the shrub.
- *Phlomis fruticosa* – requires a light annual pruning at the end of the winter. Some gardeners leave the dead flower heads on the plant as an attraction during the winter.
- *Phormium tenax* – does not require annual pruning but does benefit from a tidy up at the end of the winter. Remove any dead or damaged leaves by hand (wear gloves) or by cutting back as far as you can. Cut out any old flower stems as far down as possible without damaging the leaves. Do not attempt to reduce the height by reducing the length of the leaves.
- *Pinus mugo* – no pruning is required.
- *Prunus spinosa* – will take hard pruning if grown as part of a mixed hedge. Avoid cutting the hedge when birds are nesting. If grown as a free-standing shrub or tree, it requires minimal pruning. Cut out any damaged shoots and any crossing the centre. This will maintain a good flow of air through the plant. This pruning should be done in late summer whilst the plant is still in leaf.
- *Rhamnus alaternus* – requires minimal pruning.

Buddleja globosa has attractive, fragrant orange-yellow flowers. It should be grown in a sunny, well-drained spot.

French lavender (*Lavandula stoechas*) can be pruned in the same manner as *Lavandula angustifolia*.

- *Rosa rugosa* cultivars – can be given a light pruning in early winter. Any dead or damaged wood should be cut out and shoots that cut across the centre of the bush can also be removed. Every two or three years a few old stems can be removed to promote fresh new growth. *Rosa pimpinellifolia* cultivars can be treated in a similar fashion.
- *Rosmarinus officinalis* – removing pieces for use in the kitchen is usually enough to keep the shrub in good shape. Overgrown, woody specimens are best replaced.
- *Sambucus nigra* – requires minimal pruning but will tolerate hard pruning to reduce its size. This is best carried out in the late winter. Hard pruning will also encourage the production of larger heads of flower.
- *Santolina chamaecyparissus* – requires cutting back hard at the end of the winter. It should be cut back to where new growth is emerging. The flowering stems should be removed after flowering. An alternative approach is to remove these stems when in bud. Preventing flowering in this way produces an attractive mound of silvery foliage.
- *Sorbus* cultivars – require little or no pruning. During the winter any dead, diseased or damaged growth should be removed.
- *Spartium junceum* – will tolerate a light trim in mid spring. Formation of new shoots from old wood is not reliable and old, leggy specimens are best replaced.
- *Tamarix* cultivars – require an annual pruning but only a light one to keep the shrub under control. *Tamarix ramosissima* should be pruned at the end of winter. But *Tamarix tetrandra* should be pruned immediately after flowering and not in the winter.
- *Ulex europaeus* – requires a trim every few years to keep it tidy. This should be carried out after the main flush of flower. It is advisable to wear gloves as this is a very prickly shrub.
- *Yucca* cultivars – do not require pruning. Old flower stems and any damaged leaves can be removed in spring. If the sharp spines on the end of *Yucca gloriosa* are of concern they can be removed with a pair of nail clippers.

PLANT PROPAGATION

Many gardeners gain a great deal of satisfaction from the propagation of their plants. It is an economical way of increasing the stock of plants which can be used for maximising the planting of the garden and replacing stock that has passed its 'sell-by date'. This is particularly useful as planting looks best if several of the same plant are planted together and a good effect is achieved by repeating areas of the same plant in different parts of the garden. If a large specimen is obtained from the garden centre or nursery, it can then be used as a stock plant for propagation purposes.

Many shrubs can be propagated by taking cuttings and this is a relatively easy technique. Here are some techniques for plants typically used in coastal garden planting schemes.

Softwood cuttings

Softwood cuttings are non-flowering shoots of the current year's growth. The shoots should be pliable and therefore the cuttings are taken in spring and early summer. The cut shoot should have three to five pairs of leaves. It should be cut early in the morning and immediately placed in a plastic bag to reduce the loss of moisture through transpiration. Using a sharp knife, cut below a node (the point at which the leaf joins the stem). The cutting should be about 8–10cm (3–4 inches) long. The leaves on the lower third of the cutting should now be cut off. Insert the cutting into the compost close to the edge of a pot. A suitable compost can be made from seed compost mixed with an equal amount of perlite or horticultural grit. The pot of cuttings can now be top dressed with a layer of horticultural grit to a depth of about 1cm (0.5 inches). Leave the pot in a sheltered area such as a greenhouse or cold frame, avoiding direct sunlight as much as possible. The pot can be covered by a plastic bag with the corners cut off to ensure some flow of air. This set up will reduce loss of moisture from the cutting due to transpiration. Check the pot of cuttings daily. When the cutting shows sign of growth with fresh green foliage the cutting is starting to root. As soon as roots show through the base of the pot, remove the cuttings carefully and pot up separately into small individual pots of potting compost.

This procedure is suitable for the propagation of *Escallonia* cultivars, *Prunus spinosa*, and cultivars of *Cistus*, *Fuchsia* and *Gazania*.

Cistus can also be propagated from seed which should be sown as soon as it is ripe in containers of seed compost and left in a cold frame. Named cultivars such as *C.* 'Sunset' will need to be propagated from cuttings in order to come true to type.

Semi-hardwood cuttings

Semi-hardwood cuttings are shoots of the current season's growth. They are taken during mid to late summer and are of a similar length to the softwood cuttings described above. The cutting should be soft and flexible at the tip and firm and woody (ripe) at the base. Large leaves should have the top half removed. This will reduce the transpiration rate of the cutting. The cutting can then be prepared as outlined above. Some gardeners dip the base of the cutting in water and then rooting hormone powder; others do not. It is a matter of personal choice. The cutting can then be left in a shady spot to avoid the hot sun from causing too much transpiration. Any rotten leaves that occur should be removed. When the cuttings have rooted, they can be potted on separately.

Propagation using semi-hardwood cuttings is suitable for many shrubs that are in our coastal gardens. These include *Artemisia* 'Powis Castle', *Berberis*, *Brachyglottis*, *Buddleja* cultivars, *Convolvulus cneorum*, *Euonymus japonicus*, *Hebe*, *Hydrangea*, *Lavandula*, *Rosmarinus*, *Santolina* cultivars, and *Ulex europaeus*.

Some shrubs, e.g. *Hebe* cultivars, will root more easily with some bottom heat. Propagation units that supply bottom heat are available from garden centres. These vary in size. Soil warming cables and heat mats are also available.

These cuttings have been prepared as described. Note that the *Hydrangea* cutting has had the large leaves reduced in size.

This young plant of *Hydrangea macrophylla* 'Ayesha' was taken as a cutting the previous year. This cultivar is one of our favourite hydrangeas, in alkaline soil here; it would be a lovely shade of blue in acid soil.

Hardwood cuttings

Hardwood cuttings are taken from early autumn to early winter. Leafless shoots of fully ripe growth of the current year are used. Cut such a shoot and trim to 15–23cm (6–9 inches). Make a cut at the top above a node and at the bottom cut below a node. The position of the node will now identify the top of the cutting. Prepare a trench in the garden about four inches deep and line it with coarse sand or horticultural grit. If preferred, treat the cutting with hormone rooting powder and firm into the trench leaving 2–5cm (1–2 inches) of the cutting above the level of the soil. Signs of growth the following year will be an indication of rooting. This technique is a good method for propagating roses.

Root cuttings

Root cuttings are taken when the plant is dormant, i.e. during the winter. Several perennials can be propagated in this way. The roots should be at least 5mm (0.25 inches) in diameter and about 5–10cm (2–4 inches) long; they should be cut from close to the crown of the plant. The end of the root nearest the crown (proximal end) should be cut straight across and the opposite end cut on the diagonal. This will remind you which end of the cutting is the top. Fibrous roots should be cut away. Insert the thick root into a cutting compost with the proximal end level with the surface. Add a layer of horticultural grit over the top and leave in a cold frame, or cold greenhouse, until signs of growth appear. When in growth the rooted cutting can be potted up. This technique is suitable for *Crambe maritima* and *Eryngium* cultivars.

Eryngium × bourgatii, like all *Eryngium* cultivars, should be grown in full sun in a well-drained soil.

Division

Division is a good method of propagation for perennials with spreading rootstocks which produce new shoots on an annual basis from the crown of the plant. It is best carried out when the plants are dormant in late autumn and early spring. However, it should not be done if the conditions are very wet or frosty. If the plant is not totally hardy the division is best done in the early spring, in the hope of avoiding bad weather. When carrying out the division the plant should be lifted, and excess soil removed from it. It can then be separated into sections. This can be done by hand if the rootstock is loose. If not, use two garden forks back-to-back to lever the plant apart. If the crown of the plant is very tough, use a sharp knife or old pruning saw. The best pieces to use will be on the outer part of the crown and these can be replanted at the same depth as the original plant and watered in. The old woody centre of the plant can be discarded.

Grasses such as *Cortaderia* cultivars, *Miscanthus sinensis* and *Panicum virgatum* are best divided in the spring. *Miscanthus sinensis* cultivars may well require a saw to separate the pieces which should have a sizeable section of roots and not be very small pieces. Young plants of *Miscanthus sinensis* are more susceptible to frost damage than more established specimens, so cover with fleece if a hard frost is forecast. *Panicum virgatum* should be divided in the spring when they are starting into growth. They can also be propagated by root cuttings. Grasses can be propagated by seed but named cultivars will not come true to type. The seed should be collected when ripe and on the point of dropping from the plant. When dry, the seed can be stored in paper envelopes in a tin and left in a frost-free place. Sow the seed in spring or early summer and leave in a cold frame to germinate. The seed can be sown on the surface of a sterilised seed compost and covered with a thin layer of horticultural grit.

Yucca cultivars and *Agave americana* can be propagated by offsets. These are small plants that form around the base of the parent plant. The offsets should be removed by slicing through the stem that it is attached to the parent plant. Use a sharp spade or old pruning saw to do this, but do be careful of the spines, as they can do real damage to the skin. Usually, the offsets will have roots and can be treated as a mature plant. It there are no roots the offset should be inserted into cutting compost (*see* above) in a pot until rooted.

This fine *Agapanthus* forms a large clump that could be divided in the spring.

The 'century plant' is a common name for *Agave americana*. This is because it was thought to take a hundred years before it flowered. In reality, it takes between ten and thirty years to flower. The flower stalk can reach 5 metres (16 feet) in height. After flowering the plant dies, but do not despair for there are usually plenty of healthy offsets present.

Other plants that are likely to be in the coastal garden and have not been mentioned in this section can be propagated as indicated in Table 4.2.

TABLE 4.2 ADDITIONAL COASTAL PLANTS AND THEIR PROPAGATION METHODS

Plant	Propagation method
Ajuga reptans	by division of rooted stems
Armeria maritima	by seed or division
Crocosmia cultivars	by division in early spring
Echium vulgare	by seed on an annual basis
Elaeagnus cultivars	by semi-hardwood cuttings
Eschscholzia californica	by seed sown in spring or autumn
Foeniculum vulgare	by seed on an annual basis
Gazania cultivars	by seed or cuttings in late summer for named cultivars
Glaucium flavum	by seed in spring or autumn
Griselinia littoralis	by semi-hardwood cuttings
Kniphofia cultivars	by division in late spring
Lupinus arboreus	by seed in spring
Ranunculus ficaria 'Brazen Hussy'	by division in spring
Sisyrinchium striatum	by seed or division
Spartium junceum	by seed in spring or autumn
Trifolium repens	by seed or rooted pieces

Chapter 5

The Plants

PLANTS FOR COASTAL GARDENS

At the very heart of designing any garden is the task of choosing plants that are going to not only survive and thrive in the positions that you want to plant them but also fulfil the aesthetic or practical effect that you want to achieve with them. Not an easy task. Every plant has its own requirements in terms of the amount of sun or shade it will tolerate, the extent of winter frosts it will survive, and, particularly relevant for our purposes here, how well it will cope with salt-laden winds coming off the sea. Furthermore, you might be looking for a particular combination of colours in the plants you choose, but are your choices going to flower at the same time of year? You might have found the perfect plant in terms of its colour and tolerance for the conditions, but is it the right size for the location?

This chapter will list the plants that we have found most useful in coastal gardens and provide the details about them that you need to make the right choices. We have also included sample designs based on some common styles of planting or practical situations. You may find these helpful starting points when producing your own planting schemes. Our choice of plants as garden designers is much more limited in coastal gardens than those inland but limitations can spur our creativity and there are still lots of fantastic plants available to us.

Cordyline australis trees punctuate exuberantly colourful perennial planting in this garden designed by Andrea Rubin.

Mature heights and spreads are rough indications and will depend on the individual specimen and planting conditions.

We have indicated hardiness as follows with the RHS hardiness indicated in brackets:

- **Half hardy (RHS H3):** –5 to 1°C (23 to 34°F). Would be safest in a greenhouse during a United Kingdom winter.
- **Hardy (RHS H4):** –10 to –5°C (14 to 23°F). Hardy in most locations during a typical United Kingdom winter.
- **Very hardy (RHS H5):** –15 to –10°C (5 to 14°F). Hardy in most locations during a severe United Kingdom winter.
- **Fully hardy (RHS H6 and H7):** –15°C (–4°F) or colder. Hardy in winter throughout Northern European climates and throughout the United Kingdom in the most exposed areas during severe winter conditions.

TREES

Trees are some of the hardest plants to establish in coastal gardens. Their height means they are hard to shelter, and they must be structurally strong enough to deal with the pruning effect of strong coastal winds. So, it is worth noting that although the trees listed below are ones that we have found to be the most useful and reliable, any of them may suffer salt scorch and fail to thrive in very exposed areas. Try to plant them, where possible, in positions with some shelter, make sure you stake them well, as described in the previous chapter, and keep them well watered in the months after planting if the weather is dry.

HOW PLANTS ADAPT TO A COASTAL ENVIRONMENT

Plants that are adapted to soils high in salt or to the salt-laden winds found in coastal areas are known as halophytes, from the Ancient Greek words for salt (*halo-*) and plant (*phyte*). In addition to the salt effects, weather conditions on the coast are often windy, sunny, and mild. The combination of sun and wind increases the rate of transpiration, the process of water loss from the leaves of the plant. This water loss must be replaced by water taken from the soil by the roots of the plant, and salt in the soil affects this water uptake. Furthermore, coastal plants often need to cope with the pruning effect of strong winds and are consequently smaller and more compact in comparison with inland specimens. It is a difficult combination, and most plants are not tolerant of these conditions and will fail to thrive. Plants that can cope in coastal locations have evolved some interesting adaptations.

Examples of this include:

Hairy, thick, or scaly leaves
The effect of these adaptations is to reduce water loss. Some leaves of coastal plants look white or silver due to dense hairs, for example, the silvery leaved *Convolvulus cneorum*.

Shiny or light silvery or grey-green leaves
These reflect sunlight and heat radiation and consequently reduce desiccation. *Brachyglottis* 'Sunshine' is an example of this and makes an attractive foliage shrub if the flowers are removed when in bud.

Smaller leaves
The reduced surface area of small leaves reduces the transpiration rate. For example, the small leaved hebe cultivars, such as *Hebe* 'Wiri Cloud', are hardier than the larger-leaved cultivars which require a more sheltered position.

Waxy stems and leaves
Another adaptation that reduces the rate of transpiration. Both sea holly (*Eryngium maritimum*) and yellow horned poppy (*Glaucium flavum*) have waxy leaves and cope with the toughest of coastal conditions on shingle banks.

Long roots
Plants with long tap roots can reach water in the soil at greater depths than many other plants and consequently are less likely to suffer from water shortage. Sea kale (*Crambe maritima*) is a good example of this adaptation. They can also help give stability in constantly shifting, sandy soils. These soils are often free draining and low in fertility, and plants such as the sea holly (*Eryngium maritimum*) tend to prefer these types of soil in the garden as well.

Thick, fleshy plant parts
A lot of halophytes are succulents which store water in their leaves. *Agave americana*, with its thick, spiny, waxy leaves, is a striking example of this.

Oily foliage
Highly aromatic foliage, laden with oils, helps resist desiccation. Lavender cultivars are an example of this.

Low growth
Plants that are low growing are less exposed to the wind, reducing transpiration and damage. Sea campion (*Silene maritima*) forms a low mound of foliage and is much shorter than other campions. Some coastal plants have evolved low, prostrate stems that root as they spread, further aiding stability in windy conditions and on sandy soils.

Good examples of plants that show some of these adaptations and are very commonly seen growing wild along the British coast are the Hottentot fig (*Carpobrotus edulis*) and English stonecrop (*Sedum anglicum*).

The Hottentot fig is native to South Africa but has become naturalised in the United Kingdom, particularly along the south-west coast of England, but has reached as far as Anglesey. It is very invasive and forms large, low-growing mats of succulent leaves. It should not be planted in our gardens because of this invasive habit.

English stonecrop (*Sedum anglicum*) has small leaves and is another low-growing succulent. It is a highly successful coastal plant and is found on both dunes and shingle. It also can be a very good addition to coastal planting schemes.

A naturalised patch of the Hottentot fig, a garden escapee which is extremely invasive and a serious problem for the conservation of some coastal cliff and slope habitats as it is difficult to manage.

Alnus cordata

The Italian alder is a vigorous, medium to large tree that grows rapidly even on chalk soils. If you have an area in the garden with poor, wet soils, this tree could be a good choice as it can thrive in these conditions. It has shiny dark green leaves and produces large yellow catkins in late winter.

Mature height: 12m (40ft) or more
Mature spread: 6m (20ft)
Fully hardy
Full sun

Carpinus betulus

The common hornbeam is particularly good on both clay and chalk soils. A large, deciduous tree, it holds onto its dead leaves for much of the winter. Consequently, despite being deciduous, this is a useful tree if you need a visual screen to hide some unsightly part of the view. Hornbeam can also be used for hedging or as a pleached screen.

Mature height: 12m (40ft) or more
Mature spread: 8m (26ft) or more
Fully hardy
Full sun/partial shade

Carpinus betulus 'Fastigiata'

This cultivar of the common hornbeam has an erect, pyramidal habit and is well worth consideration for growing within a hedge but also as a tree, adding architectural effect, selective screening, or perhaps framing a view.

Mature height: 12m (40ft) or more
Mature spread: 3m (10ft)
Fully hardy
Full sun/partial shade

Carpinus betulus 'Columnaris'

This cultivar is a smaller, bushier, slower-growing version of the common hornbeam. It is especially useful for the smaller garden.

Mature height: 6m (20ft) or more
Mature spread: 3m (10ft)
Fully hardy
Full sun/partial shade

Cordyline australis

The cabbage palm is a small evergreen tree with panicles of fragrant flowers followed by small berries. It is hardy in relatively mild areas, particularly coastal ones where it can be an excellent feature and a useful tree in planting schemes. It can be badly affected, if not killed, by hard frosts. In very exposed areas the leaves can be damaged by strong winds.

Mature height: 7m (23ft)
Mature spread: 5m (17ft)
Half hardy
Full sun/partial shade

Crataegus monogyna

Common hawthorn, very useful for hedging and, grown as a tree, is a spectacular sight when in full flower in May. The flowers are strongly fragrant and followed by red fruit. Once established it is tolerant of both dryness and excessive moisture.

Mature height: 6m (20ft) or more
Mature spread: 6m (20ft)
Fully hardy
Full sun/partial shade

The cabbage palm (*Cordyline australis*).

Crataegus monogyna 'Stricta'

A fastigiate, columnar form of the common hawthorn that is an excellent, small tough cultivar which, grown as a tree, is also useful for its architectural effect, as screening, or where space is limited.

Mature height: 6m (20ft) or more
Mature spread: 3m (10ft)
Fully hardy
Full sun/partial shade

Crataegus monogyna 'Stricta', a fastigiate form of hawthorn.

Crataegus laevigata

The Midland hawthorn is native to the United Kingdom like *C. monogyna* and produces an equally good show of flowers in May. As a tree, it has a small, low branching habit. Common in hedgerows. *C. laevigata* 'Crimson Cloud' has single crimson flowers with white centres.

Mature height: 5m (16ft)
Mature spread: 5m (16ft)
Fully hardy
Full sun/partial shade

Crataegus laevigata 'Paul's Scarlet' is a double-flowered form of the Midland hawthorn.

Crataegus laevigata 'Aurea' is a rare form with yellow fruits.

Cupressus macrocarpa

This fast-growing large conifer, the Monterey cypress, provides excellent shelter from coastal winds, if you have the room for it. Could be used for hedging but would require very regular pruning. Suitable only for large gardens.

Mature height: 12m (40ft) or more
Mature spread: 8m (26ft) or more (becomes less columnar, more wide spreading with age)
Fully hardy
Full sun/partial shade

Cupressus macrocarpa 'Goldcrest'

A columnar form of the Monterey cypress which is both dense and compact. Fantastic, bright yellow colour but can scorch in very exposed sites. Can be particularly good as a hedge.

Mature height: 12m (40ft) or more
Mature spread: 6m (20ft)
Fully hardy
Full sun/partial shade

Eucryphia × nymansensis 'Nymansay'

This is a columnar tree which is hardy in southern parts of the United Kingdom. It will require some shelter from the coastal winds and would be very happy in light woodland and on acidic or neutral soil. It is a magnificent sight in latish summer, its large white flowers being very attractive to both bees and hoverflies.

Mature height: 15m (49ft)
Mature spread: 5m (16ft)
Hardy
Full sun/partial shade

Phillyrea latifolia

A small tree or large evergreen shrub with fragrant but small flowers. A useful substitute where *Quercus ilex* would be too large.

Mature height: 7m (23ft)
Mature spread: 7m (23ft)
Very hardy
Full sun/partial shade

Eucryphia × nymansensis 'Nymansay'.

Pinus contorta

A large conifer, the lodgepole pine is unsuitable for chalk soils but does well on sandy ones. Can be an interesting shape when mature and has attractive bark.

Mature height: 12m (40ft) or more
Mature spread: 8m (26ft) or more
Very hardy
Full sun/partial shade

Pinus nigra

The Austrian pine is an excellent conifer for maritime areas and is the best pine for chalky soils.

Mature height: 12m (40ft) or more
Mature spread: 8m (26ft) or more
Very hardy
Full sun/partial shade

Pinus nigra 'Maritima'

This Corsican pine has long slightly twisted spines. It is happy on most soils.

Mature height: 12m (40ft) or more
Mature spread: 8m (26ft) or more
Very hardy
Full sun

Pinus radiata

The Monterey pine is a large, fast-growing conifer which has an excellent capability for withstanding sea winds.

Mature height: 12m (40ft) or more
Mature spread: 8m (26ft) or more
Very hardy
Full sun

Quercus ilex

The evergreen 'holm oak' is a common tree in many coastal situations but, as a tree, is only suitable for large gardens. It thrives on any well-drained soil, tolerates shade, and responds well to clipping. *Quercus ilex* makes an excellent windbreak in large coastal gardens.

Mature height: 12m (40ft) or more
Mature spread: 8m (26ft) or more
Hardy
Full sun

Salix alba

A British native willow species with silvery leaves and slender catkins. It is a fast-growing tree that is much planted in damp sandy areas in coastal regions. Whether grown as a tree or a shrub it should be planted far away from drains. A useful addition to your planting plan if you have a damp area to deal with.

Mature height: 12m (40ft) or more
Mature spread: 8m (26ft) or more
Very hardy
Full sun

S. alba var. *sericea* is a less vigorous form of *Salix alba* with very silvery leaves.

Salix alba, the white willow.

Sorbus aria

This deciduous tree, the whitebeam, is often recommended as one of the best trees for coastal gardens but in very exposed areas it can still suffer from scorch. It is happy on chalk and has silvery grey foliage turning to gold in autumn which is accompanied by deep crimson berries.

Mature height: 12m (40ft) or more
Mature spread: 8m (26ft) or more
Very hardy
Full sun

S. aria 'Lutescens' has particularly striking foliage that looks especially good in spring. An attractive tree for coastal planting given a little shelter from the prevailing winds.

Sorbus aria 'Lutescens'.

Sorbus aucuparia

The rowan is an easily grown British native species but is not long lived on shallow chalky soils. It produces good crops of red berries in late summer which are soon devoured by blackbirds and thrushes. Cultivars with different coloured berries keep their fruit for longer. It is a feature in the garden in both spring and autumn.

Mature height: 12m (40ft) or more
Mature spread: 8m (26ft) or more
Very hardy
Full sun

S. aucuparia 'Fructu Luteo' has yellowish fruit.

S. aucuparia 'Sheerwater Seedling' is a particularly good cultivar with orange-red fruit.

The rowan (*Sorbus aucuparia*).

LARGE SHRUBS

Some of the larger shrubs can be grown as small trees, if so required, encouraging a leading shoot by removing lower competing branches.

Buddleja globosa

This *Buddleja* has globular orange-yellow flowers in summer which are totally different in shape to the more common *Buddleja davidii*. In milder areas it is semi-evergreen but perhaps requires a little shelter from the prevailing salt winds.

Mature height: 5m (15ft)
Mature spread: 5m (15ft)
Very hardy
Full sun

***Buddleja* × *weyeriana* 'Sungold'**

With fragrant, conical orange-yellow flowers in summer to autumn, this shrub attracts bees and other pollinators. It grows best in well-drained soil in full sun. It is best pruned after flowering. May require some shelter from the worst of coastal winds.

Mature height: 4m (12ft)
Mature spread: 3m (9ft)
Very hardy
Full sun

Elaeagnus angustifolia

Elaeagnus angustifolia is a large, deciduous, thorny shrub with grey leaves and fragrant flowers in June. It can be grown as a small tree. The flowers are followed by 2.5cm long yellowish berries. Its leaves resemble those of a willow. It can withstand temperatures down to about –15°C (5°F).

Mature height: 6m (20ft)
Mature spread: 6m (20ft)
Fully hardy
Full sun

Elaeagnus × ebbingei

Elaeagnus × ebbingei grows well on any fertile soil except shallow chalk. It is a large evergreen shrub which has lots of small, scented flowers in autumn. It makes an excellent hedge as it is fast growing and responds well to regular pruning.

Mature height: 4m (12ft)
Mature spread: 4m (12ft)
Fully hardy
Full sun/partial shade

E. × ebbingei 'Gilt Edge' has leaves edged golden yellow.

Elaeagnus × ebbingei 'Limelight'

E. × ebbingei 'Limelight' has leaves with a central splash of pale green and yellow. It does tend to revert, i.e., produce all green shoots. These should be completely cut out or they will predominate.

Mature height: 3m (10ft)
Mature spread: 3m (10ft)
Fully hardy
Full sun/partial shade

Elaeagnus × ebbingei.

Elaeagnus pungens

This *Elaeagnus* is a vigorous evergreen with fragrant silvery white flowers in autumn.

Mature height: 4m (12ft)
Mature spread: 5m (15ft)
Fully hardy
Full sun/partial shade

E. pungens 'Maculata' is a vigorous form. Its leaves have a central splash of gold. It is a handsome shrub but with a tendency to reversion.

Elaeagnus pungens 'Frederici'

E. pungens 'Frederici' is a slow-growing form with creamy yellow variegation.

Mature height: 2m (6ft)
Mature spread: 2m (6ft)
Fully hardy
Full sun/partial shade

Elaeagnus 'Quicksilver'

E. 'Quicksilver' is a large deciduous shrub resembling *E. angustifolia*. The foliage is very silvery. The plant tends to sucker.

Mature height: 4m (12ft)
Mature spread: 4m (12ft)
Fully hardy
Full sun

Elaeagnus pungens 'Frederici'.

Escallonia rubra var. *macrantha*

This *Escallonia* will grow as high as 3m and has crimson flowers and glossy aromatic leaves. Very tolerant of coastal conditions, it is one of the best coastal hedging plants, flowering well, fast growing, and happy with regular pruning.

Mature height: 3m (10ft)
Mature spread: 3m (10ft)
Fully hardy
Full sun

Griselinia littoralis

Griselinia makes an excellent evergreen hedge in maritime situations. It may become tree-like in mild situations but is liable to frost damage in colder areas. The flowers are inconspicuous.

Mature height: 8m (25ft)
Mature spread: 5m (15ft)
Hardy
Full sun

Griselinia littoralis 'Dixon's Cream'

G. littoralis 'Dixon's Cream' has leaves that are splashed and marked creamy white. It is slower growing than the all-green form, so do not mix them.

Mature height: 3m (10ft)
Mature spread: 2m (6ft)
Hardy
Full sun

Escallonia rubra var. *macrantha*.

Hippophae rhamnoides

Sea buckthorn is a silver-leaved native shrub or small tree. It is easy to grow, very tolerant of coastal conditions, and produces orange berries on the female plant.

Mature height: 6m (20ft)
Mature spread: 6m (20ft)
Fully hardy
Full sun

H. rhamnoides 'Pollmix' is a male form which can be selected as a pollinator.

Olearia traversii

This *Olearia* is one of the best and fastest growing evergreens for milder coastal areas. It is hardy down to about –5°C (23°F). It can reach up to 6m in height but is also a particularly good hedging plant.

Mature height: 6m (20ft)
Mature spread: 4m (12ft)
Hardy
Full sun

Hippophae rhamnoides.

Phillyrea angustifolia

It forms a compact rounded evergreen bush and is excellent for coastal areas. The flowers are insignificant but fragrant.

Mature height: 3m (10ft)
Mature spread: 3m (10ft)
Very hardy
Full sun

***Phormium* 'Bronze Baby'**

Phormium 'Bronze Baby' is a small cultivar with bronze leaves and can be used to shelter smaller plants. A planting of *Gazania* in front of it is very effective, as are sedums or lavender.

Mature height: 0.8m (2.5ft)
Mature spread: 0.8m (2.5ft)
Hardy
Full sun

Phillyrea angustifolia.

Phormium tenax

Phormium tenax is a fine architectural plant which makes a bold statement in a planting scheme. It has red panicles of flowers in summer. In very exposed positions the leaf edges may get frayed. Plain coloured varieties are more wind tolerant than variegated cultivars.

Mature height: 4m (12ft)
Mature spread: 2m (6ft)
Hardy
Full sun

P. tenax 'Purpureum' is a vigorous grower with purple leaves.

***Phormium tenax* 'Platt's Black'**

Phormium tenax 'Platt's Black' is another small cultivar. Its size means that it would work well planted in a container.

Mature height: 1.2m (4ft)
Mature spread: 1.2m (4ft)
Hardy
Full sun

Pinus mugo

This pine makes a large shrub or a small tree. It is very lime tolerant and there are several dwarf forms such as 'Gnom', 'Ophir', and 'Winter Gold' which can be used for screening more sensitive plants.

Mature height: 3.5m (11ft)
Mature spread: 5m (15ft)
Fully hardy
Full sun

Phormium tenax 'Purpureum'.

Prunus spinosa

Blackthorn is a British native large shrub or small tree which makes an excellent hedge and is very attractive to wildlife. It produces masses of white flowers in spring which are followed by black fruit which can be used to flavour gin. Avoid planting over drains and close to foundations of buildings.

Mature height: 5m (15ft)
Mature spread: 4m (12ft)
Fully hardy
Full sun

P. spinosa 'Plena' has double flowers and makes an excellent small tree.

P. spinosa 'Purpurea' is one of the best purple-leafed shrubs.

Rhamnus alaternus

Italian buckthorn is a large, fast-growing evergreen. It has yellowish green flowers in spring which are followed by red fruit which then turn black. It is not suitable for very cold areas.

Mature height: 5m (15ft)
Mature spread: 4m (12ft)
Hardy
Full sun/partial shade

The blackthorn (*Prunus spinosa*).

Sambucus nigra

Elder is a British native and is a large shrub or small tree which grows well on all soils including very chalky ones. It has flat heads of fragrant white flowers which are followed by bunches of black berries. Hard pruning in spring will help to produce large heads of flowers.

Mature height: 6m (20ft)
Mature spread: 6m (20ft)
Fully hardy
Full sun/partial shade

S. nigra 'Black Lace' has excellent dark purple foliage as well as flowers and berries.

S. nigra 'Fructu Luteo' has yellow fruits.

S. nigra 'Laciniata' has very attractive deeply dissected fern-like leaves and cream flowers.

Sambucus nigra 'Black Lace'.

Sambucus nigra 'Laciniata'.

Tamarix ramosissima

It is a large shrub or small tree which thrives in any soil except for shallow chalk ones and is extremely tolerant of coastal exposure. Its flowers, on the current year's growth, are pink. It should be pruned at the end of winter.

Mature height: 5m (15ft)
Mature spread: 5m (15ft)
Fully hardy
Full sun

T. ramosissima 'Pink Cascade' is an excellent cultivar with large plumes of pink flowers in July and August.

T. ramosissima 'Rosea' flowers in late summer and early autumn and is one of the finest late-flowering shrubs.

Tamarix tetrandra

This *Tamarix* flowers in May and June on growth made the previous year. It should be pruned immediately after flowering.

Mature height: 3m (10ft)
Mature spread: 3m (10ft)
Fully hardy
Full sun

MEDIUM SHRUBS

Artemisia arborescens 'Powis Castle'

This *Artemisia* is an evergreen with attractive mounds of aromatic filigree silver foliage. It should be grown in full sun in a well-drained soil. It should be pruned annually to keep it compact.

Mature height: 0.6m (2ft)
Mature spread: 0.9m (3ft)
Hardy
Full sun

Atriplex halimus

This shrub, a true halophyte, thrives on salty soils. Its flowers are inconspicuous, and it is grown for its silvery grey leaves. In cold areas it is partially deciduous. It makes an excellent informal hedge.

Mature height: 2m (6ft)
Mature spread: 2.5m (8ft)
Hardy
Full sun

Berberis darwinii

This is one of the finest of flowering shrubs. It flowers in spring and is evergreen with small holly-like leaves and orange flowers.

Mature height: 1.5m (5ft)
Mature spread: 1.5m (5ft)
Fully hardy
Full sun/partial shade

Tamarix ramosissima 'Pink Cascade'.

Artemisia 'Powis Castle'.

Berberis julianae

This *Berberis* is also evergreen and makes a good hedge up to 3m high. It produces yellow or red tinged flowers in late spring, and these are followed by black fruit.

Mature height: 3m (10ft)
Mature spread: 3m (10ft)
Fully hardy
Full sun/partial shade

Berberis × ottawensis

This *Berberis* is deciduous and is a first-class tall, arching shrub with purple leaves and pale-yellow flowers which are followed by red berries. Like all *Berberis* it flowers and fruits better in full sun.

Mature height: 2.5m (8ft)
Mature spread: 2.5m (8ft)
Fully hardy
Full sun/partial shade

Atriplex halimus.

Berberis thunbergii

Berberis thunbergii is a dense, rounded deciduous green-leaved shrub with brilliant autumn colour. It is excellent for hedging. There are several excellent cultivars.

Mature height: 1m (3ft)
Mature spread: 2.5m (8ft)
Fully hardy
Full sun/partial shade

B. thunbergii f. *atropurpurea* is a deciduous *Berberis* with excellent autumn leaf colour.

B. thunbergii 'Dart's Red Lady' has deep purple leaves which turn a brilliant red in autumn.

Berberis darwinii.

Berberis thunbergii f. *atropurpurea.*

***Berberis thunbergii* 'Diabolicum'**

This is a compact form with yellow leaves and orange-red edges in spring. Insignificant yellow flowers in spring are followed by bright red fruit.

Mature height: 0.5m (2ft)
Mature spread: 0.5m (2ft)
Fully hardy
Full sun/partial shade

***Berberis thunbergii* 'Golden Torch'**

'Golden Torch' is a slow-growing shrub with an upright habit. It has white flowers followed by red fruit plus good autumn-coloured foliage. It works well in combination with purple-leaved cultivars.

Mature height: 1.5m (5ft)
Mature spread: 1m (3ft)
Fully hardy
Full sun/partial shade

Berberis thunbergii 'Golden Torch'.

***Brachyglottis* 'Sunshine'**

This shrub was formerly known as *Senecio* 'Sunshine'. It forms a broad mound of silvery grey leaves with yellow flowers. It is a sun-loving evergreen for well-drained soils.

Mature height: 1.5m (5ft)
Mature spread: 2m (6ft)
Hardy
Full sun

***Brachyglottis* 'Drysdale'**

'Drysdale' is a more compact form of this shrub.

Mature height: 0.8m (2.5ft)
Mature spread: 0.8m (2.5ft)
Hardy
Full sun

Brachyglottis 'Silver Dormouse' is another good compact form with similar dimensions to 'Drysdale'.

Bupleurum fruticosum

This is one of the best medium-sized evergreen shrubs for use near the sea. It has umbels of yellow flowers from midsummer to early autumn.

Mature height: 2m (6ft)
Mature spread: 2.5m (8ft)
Hardy
Full sun

Brachyglottis 'Sunshine'.

Ceanothus cultivars

Ceanothus prefer full sun and well-drained soil. Many cultivars require some shelter rather than complete exposure to salt winds. Evergreen cultivars should be pruned after flowering whilst deciduous cultivars should be pruned in early spring. Most can be trained against a wall; they will reach a greater height here than when grown in the open.

Ceanothus arboreus 'Trewithen Blue'

'Trewithen Blue' is evergreen and a vigorous grower with large deep-blue flowers in spring and early summer.

Mature height: 6m (20ft)
Mature spread: 8m (25ft)
Hardy
Full sun

Ceanothus 'Gloire de Versailles'

'Gloire de Versailles' is a deciduous cultivar with large powder-blue flowers in summer and autumn.

Mature height: 1.5m (5ft)
Mature spread: 1.5m (5ft)
Very hardy
Full sun

Bupleurum fruticosum.

Ceanothus 'Puget Blue'

'Puget Blue' is an evergreen cultivar with outstanding deep blue flowers in late spring and early summer.

Mature height: 2m (6ft)
Mature spread: 2m (6ft)
Very hardy
Full sun

Ceanothus thyrsiflorus var. *repens*

This cultivar is a low grower and has a wide spreading habit and light blue flowers in June. It is evergreen.

Mature height: 1m (3ft)
Mature spread: 2.5m (8ft)
Hardy
Full sun

Ceanothus 'Puget Blue'.

Chamaerops humilis

This is our first choice of palm for coastal planting. It rarely exceeds 2m (6ft) high and is hardy in mild localities. It produces suckers when mature which can be removed for propagation or to limit its spread but should be left on if possible.

Mature height: 2m (6ft)
Mature spread: 1.5m (5ft)
Hardy
Full sun/partial shade

Cistus cultivars

Cistus are bushy evergreens that require full sun and a well-drained soil. They do not respond well to pruning and may be killed in severe winters. Young plants should be pinched back after flowering to produce a bushy habit.

Cistus × *aguilarii* 'Maculatus'

This cultivar has large white flowers with central crimson blotches.

Mature height: 1.2m (4ft)
Mature spread: 1.2m (4ft)
Hardy
Full sun

Chamaerops humilis.

Cistus × *dansereaui* 'Decumbens' syn. *Cistus lusitanica* 'Decumbens'

This cultivar has white flowers with a maroon blotch. It is one of the hardier rock roses but will require some shelter from the worst of the prevailing winds

Mature height: 1m (3ft)
Mature spread: 1.2m (4ft)
Hardy
Full sun

Cistus × hybridus syn. *Cistus corbariensis*

This cultivar has white flowers with yellow centres in early summer. It does require some shelter from the prevailing wind.

Mature height: 0.9m (3ft)
Mature spread: 0.9m (3ft)
Hardy
Full sun

Cistus × *purpureus*

This is a rounded shrub and in summer has rose-pink flowers with central chocolate blotches.

Mature height: 1m (3ft)
Mature spread: 1m (3ft)
Hardy
Full sun

C. × *purpureus* 'Alan Fradd' has large white flowers with chocolate blotches. It often shows signs of partial reversion within the flower.

Cistus × *purpureus* 'Alan Fradd'.

Cistus 'Silver Pink'

This cultivar has clusters of soft pink flowers and is more hardy than other cultivars.

Mature height: 0.75m (2.5ft)
Mature spread: 0.9m (3ft)
Hardy
Full sun

Convolvulus cneorum

This *Convolvulus* is a small, silver-leaved evergreen which requires well-drained soil and full sun. It has a long flowering season with white flowers.

Mature height: 0.6m (2ft)
Mature spread: 0.9m (3ft)
Hardy
Full sun

Convolvulus cneorum.

Convolvulus cneorum 'Snow Angel'

This cultivar has very silvery leaves and is shorter and bushier than the type plant.

Mature height: 0.5m (1.5ft)
Mature spread: 0.5m (1.5ft)
Hardy
Full sun

Cytisus scoparius

Common broom is an erect, deciduous shrub which has masses of yellow flowers in late spring. When pruning, do not cut into old wood.

Mature height: 1.5m (5ft)
Mature spread: 1.5m (5ft)
Fully hardy
Full sun

Escallonia 'Apple Blossom'

This is a summer-flowering evergreen with pink and white flowers.

Mature height: 2.5m (8ft)
Mature spread: 2.5m (8ft)
Hardy
Full sun

Escallonia 'C.F. Ball' is a variety with crimson flowers. It makes a good hedging plant.

Escallonia 'Iveyi'

Escallonia 'Iveyi' has white flowers during the summer. It does not require pruning. If it must be done, pruning in spring will result in no flowers for that year.

Mature height: 2.5m (8ft)
Mature spread: 2.5m (8ft)
Hardy
Full sun/partial shade

Euonymus fortunei 'Emerald 'n' Gold'

This dwarf evergreen shrub is one of several cultivars of *E. fortunei* with bright golden variegated leaves which turn pink in winter.

Mature height: 0.5–1m (2–3ft)
Mature spread: 1–1.5m (3–5ft)
Hardy
Sun/partial shade

Euonymus japonicus

Euonymus japonicus is a densely branched evergreen shrub which can be used for hedging. Variegated cultivars require sun to enhance the variegation.

Mature height: 4m (12ft)
Mature spread: 2m (6ft)
Hardy
Sun/partial shade

E. japonicus 'Albomarginatus' has rounded leaves with white margins.

E. japonicus 'Ovatus Aureus' is both bushy and compact. The leaves have creamy yellow margins. It requires full sun for good colour.

Euonymus japonicus 'Aureopictus'

This variegated cultivar has elliptical leaves with bright yellow centres. It is liable to revert to its non-variegated form.

Mature height: 1.5m (5ft)
Mature spread: 1m (3ft)
Hardy
Sun/partial shade

Euonymus japonicus 'Président Gauthier'

This cultivar has dark centred leaves with a creamy white edge.

Mature height: 2m (6ft)
Mature spread: 2m (6ft)
Hardy
Sun/partial shade

Fuchsia magellanica

This hardy fuchsia is a graceful shrub of medium size. It is free flowering with long, slender scarlet and violet flowers. It thrives in sun or shade. At the end of the winter, it can be cut to the base in frost-prone areas. It is suitable for hedging in frost-free areas.

Mature height: 3m (10ft)
Mature spread: 2.5m (8ft)
Hardy
Sun/partial shade

Fuchsia magellanica 'Hawkshead'

In this cultivar there are single white flowers tinged green.

Mature height: 1m (3ft)
Mature spread: 0.5m (2ft)
Hardy
Sun/partial shade

Euonymus fortunei 'Emerald 'n' Gold'.

Fuchsia magellanica.

Fuchsia magellanica 'Sharpitor'

Fuchsia 'Sharpitor' has decorative leaves which are grey-green and white.

Mature height: 1m (3ft)
Mature spread: 1m (3ft)
Hardy
Sun/partial shade

Fuchsia magellanica 'Sharpitor'.

Fuchsia magellanica var. molinae

This cultivar has white flowers which are tinged pale pink.

Mature height: 2.5m (8ft)
Mature spread: 1.5m (5ft)
Hardy
Sun/partial shade

Genista hispanica.

Genista hispanica

This Spanish gorse is deciduous, forms a dense prickly mound with masses of yellow flowers in late spring and early summer, and is very able to cope with coastal conditions. It requires a sunny, well-drained position.

Mature height: 0.75m (2.5ft)
Mature spread: 1.5m (5ft)
Hardy
Full sun

Genista lydia

This is a dwarf, deciduous shrub with clusters of yellow flowers in early summer. It is an excellent plant for rockeries, or for trailing over low walls or the edge of raised beds.

Mature height: 0.5m (1.5ft)
Mature spread: 1m (3ft)
Hardy
Full sun

Genista lydia.

Hebe × franciscana 'Blue Gem'.

Hebe × franciscana 'Blue Gem'

Several hebe cultivars can cope with maritime conditions. *Hebe* 'Blue Gem' is one of the hardiest and is a familiar sight in coastal towns. It is an excellent choice for low hedges and produces violet-blue flowers from midsummer onwards.

Mature height: 1.3m (4.5ft)
Mature spread: 1.3m (4.5ft)
Hardy
Sun/partial shade

Hebe salicifolia

This willow-leaved hebe with spikes of white or pale lilac flowers is suitable for coastal gardens but will need some shelter from the worst of conditions.

Mature height: 5m (15ft)
Mature spread: 5m (15ft)
Hardy
Sun/partial shade

Hebe salicifolia.

Hebe 'Wiri Cloud'

This hebe is a dwarf shrub which has attractive white flowers flushed with pink.

Mature height: 0.5m (2ft)
Mature spread: 0.5m (2ft)
Hardy
Sun/partial shade

Hebe 'Wiri Image'

This is another dwarf hebe but with violet flowers.

Mature height: 1m (3ft)
Mature spread: 0.75m (2.5ft)
Hardy
Sun/partial shade

Hebe 'Wiri Cloud'.

Hydrangea macrophylla

This group of hydrangeas comes in two types: mopheads and lacecaps. The mopheads have large, almost spherical heads consisting of sterile florets. The lacecaps have circles of tiny flowers surrounded by sterile florets. They perform well in partial shade, as do all hydrangeas.

Hydrangea macrophylla – lacecap.

***Hydrangea macrophylla* 'Lanarth White'**

This is a lacecap hydrangea and is a reliable, medium-sized shrub with dark blue flowers and white florets. It is good in partial shade but will tolerate full sun. It does not like dryness at the roots.

Mature height: 1.5m (5ft)
Mature spread: 1.5m (5ft)
Fully hardy
Sun/partial shade

***Hydrangea macrophylla* 'Joseph Banks'**

This is a mophead hydrangea with large flower heads which are cream at first, fading to pale pink or blue. It will be blue on acidic soils and pink on alkaline ones.

Mature height: 2m (6ft)
Mature spread: 2.5m (8ft)
Fully hardy
Sun/partial shade

Hydrangea macrophylla – mophead.

***Hydrangea macrophylla* 'La Marne'**

H. 'La Marne' is also a mophead with large flower heads of pink or blue. It is an excellent cultivar for the coast.

Mature height: 1.5m (5ft)
Mature spread: 1.5m (5ft)
Fully hardy
Sun/partial shade

***Hydrangea macrophylla* 'Mathilda Gutges'**

This is another mophead cultivar with pink or violet-blue flowers and performs well in coastal gardens.

Mature height: 1.5m (5ft)
Mature spread: 1.5m (5ft)
Fully hardy
Sun/partial shade

Hydrangea 'Mathilda Gutges'.

Lavandula angustifolia

Lavender likes a well-drained soil in full sun. It blooms from mid- to late summer with dense spikes of pale to deep purple flowers. Old stems should be pruned at the end of winter but avoid cutting below the previous year's growth. In very exposed coastal areas, it is likely to suffer from scorching and may not survive.

Mature height: 1m (3ft)
Mature spread: 1.2m (4ft)
Fully hardy
Full sun

Lavandula angustifolia 'Hidcote'

This cultivar is more compact and has dark purple flowers.

Mature height: 0.6m (2ft)
Mature spread: 0.75m (2.5ft)
Fully hardy
Full sun

Lavandula angustifolia 'Munstead'

This is another recommended cultivar and has blue-purple flowers.

Mature height: 0.45m (1.5ft)
Mature spread: 0.6m (2ft)
Fully hardy
Full sun

Lupinus arboreus

This is a semi-evergreen, shrubby lupin. It has lots of spikes of fragrant yellow flowers in late spring and summer. It is not long lived but will self-seed. It is happiest on sandy soils. *Lupinus arboreus* has naturalised in the United Kingdom, particularly in sand dunes, where it has become a threat to native plants in some areas such as Hampshire and Norfolk. So, care should be taken not to allow it to spread seed to vulnerable wild areas.

Mature height: 2m (6ft)
Mature spread: 2m (6ft)
Hardy
Full sun

L. arboreus 'Blue Boy' is similar but with scented blue flowers.

L. arboreus 'Snow Queen' is also similar but with scented white flowers.

Olearia cultivars

Olearia cultivars include some of the finest of all evergreens for coastal exposure. They are recommended for chalky soils. Some may be grown as hedges and windbreaks in coastal areas.

Olearia × *haastii*

Olearia × *haastii* makes an excellent hedging plant and has masses of scented white flowers in summer.

Mature height: 2m (6ft)
Mature spread: 3m (10ft)
Fully hardy
Full sun

Lavandula angustifolia.

Lupinus arboreus growing in the wild.

Olearia macrodonta

This *Olearia* is a vigorous, upright shrub or small tree and has panicles of fragrant flowers in early summer. It is one of the best screening or hedging plants.

Mature height: 6m (20ft)
Mature spread: 5m (15ft)
Fully hardy
Full sun

Olearia macrodonta.

Olearia × scilloniensis

This is a very free-flowering hybrid but is only suitable for milder regions.

Mature height: 2m (6ft)
Mature spread: 2m (6ft)
Hardy
Full sun

Olearia × scilloniensis.

Phlomis fruticosa

Jerusalem sage is evergreen and requires a sunny, well-drained position. It has whorls of yellow flowers over a long period in summer. It copes very well with coastal conditions.

Mature height: 0.5m (1.5ft)
Mature spread: 1.5m (5ft)
Very hardy
Full sun

Pinus mugo 'Mops'

'Mops' is a slow-growing cultivar of the dwarf mountain pine. It grows about 7.5cm (3 inches) a year.

Mature height: 1m (3ft)
Mature spread: 1m (3ft)
Fully hardy
Full sun

A young *Phlomis fruticosa.*

Rosa pimpinellifolia

The burnet rose is a small, British native suckering rose and is very prickly. It flowers in May and June and does not repeat flower. The flowers are single, creamy white and are followed by purplish black hips.

Mature height: 1m (3ft)
Mature spread: 1.2m (4ft)
Fully hardy
Full sun

R. pimpinellifolia 'Burnet Double White' has flowers which are produced on arching branches and are followed by small blackish hips in late summer.

Rosa 'Stanwell Perpetual' is a burnet rose which blooms all summer with very fragrant blush-pink flowers.

***Rosa pimpinellifolia* 'Dunwich Rose'**

This is a low-growing cultivar with white flowers.

Mature height: 0.65m (2ft)
Mature spread: 1.2m (4ft)
Fully hardy
Full sun

Rosa rugosa

This rose is probably the healthiest of all roses. Both double and single cultivars are highly scented, and several cultivars bear large ornamental hips. The species has pink flowers followed by round red hips. It suckers quite freely, making a superb hedge.

Mature height: 2m (6ft)
Mature spread: 2m (6ft)
Fully hardy
Full sun

R. rugosa 'Alba' is a white form in which the late flowers contrast well with the red hips. It is also excellent for hedging. It should not be dead headed, as this would remove the hips which are an attractive feature in late summer.

***Rosa rugosa* 'Blanc Double de Coubert'**

This rugosa rose has large, semi-double white flowers. It repeats and has occasional hips.

Mature height: 1.5m (5ft)
Mature spread: 1.2m (4ft)
Fully hardy
Full sun

Rosa 'Stanwell Perpetual'.

Rosa rugosa.

Rosa rugosa 'Fru Dagmar Hastrup'

This rugosa rose has single, pale pink, flowers. These are followed by good autumn leaf colour and crimson hips.

Mature height: 1m (3ft)
Mature spread: 1.2m (4ft)
Fully hardy
Full sun

Rosa rugosa 'Roseraie de l'Hay'

This rugosa rose has purple, crimson flowers with a good perfume. It has good autumn leaf colour and makes an excellent hedge.

Mature height: 2.2m (7ft)
Mature spread: 2m (6ft)
Fully hardy
Full sun

Rosa rugosa 'Scabrosa'

This rose has large, deep pinkish-mauve flowers which are followed by huge red hips.

Mature height: 1.7m (6ft)
Mature spread: 1.7m (6ft)
Fully hardy
Full sun

Rosmarinus officinalis

Rosemary forms a dense shrub. It bears clusters of blue flowers along the branches of the previous year's growth. It can make an attractive informal hedge. If needed, prune lightly after flowering.

Mature height: 1.5m (5ft)
Mature spread: 1.5m (5ft)
Hardy
Full sun

Rosmarinus officinalis 'Miss Jessopp's Upright'

This is a strong-growing cultivar with an upright habit. It is a good choice for hedging.

Mature height: 2m (6ft)
Mature spread: 2m (6ft)
Hardy
Full sun

Rosmarinus officinalis 'Prostratus'

This is the least hardy cultivar and has a dense prostrate habit growing into large mats with mid-blue flowers.

Mature height: 0.2m (0.5ft)
Mature spread: 1.5m (5ft)
Hardy
Full sun

Rosmarinus officinalis 'Tuscan Blue'

This is a good cultivar with deep blue flowers. It often flowers in the winter.

Mature height: 1.5m (5ft)
Mature spread: 1.5m (5ft)
Hardy
Full sun

Rosmarinus officinalis 'Prostratus' self-seeded into and growing well in a harbour wall.

Santolina chamaecyparissus.

Santolina chamaecyparissus

Cotton lavender forms low-growing mounds with finely divided foliage. This species has silvery foliage with lemon-yellow flowers. If the flower buds are trimmed off before the flowers open the mound of foliage stays denser and more compact and makes an attractive feature in itself.

Mature height: 0.5m (1.5ft)
Mature spread: 1m (3ft)
Hardy
Full sun

Santolina chamaecyparissus 'Nana'

This cultivar is a more compact variety.

Mature height: 0.25m (1ft)
Mature spread: 0.5m (1.5ft)
Hardy
Full sun

Santolina rosmarinifolia.

Santolina rosmarinifolia

This *Santolina* has thread-like vivid green leaves with lemon-yellow flowers. The foliage has an attractive fragrance.

Mature height: 0.6m (2ft)
Mature spread: 1m (3ft)
Hardy
Full sun

S. rosmarinifolia 'Primrose Gem' is an attractive form with pale yellow flowers.

Spartium junceum.

Spartium junceum

An upright shrub which can grow to about 3m in height, and copes well with coastal winds. It grows particularly well on chalk soils. It has masses of yellow scented flowers in summer. It does not respond well to being moved. Any straggly growths can be cut out in spring.

Mature height: 2.5–3m (8–9ft)
Mature spread: 2m (6ft)
Hardy
Full sun

Ulex europaeus

This gorse, native to the UK, is excellent for poor, dry, acidic soils, and grows extremely well in exposed coastal conditions. The main flowering season is in spring, but it flowers intermittently throughout the year.

Mature height: 0.6m (2ft)
Mature spread: 1m (3ft)
Fully hardy
Full sun

U. europaeus 'Flore Pleno' is an outstanding compact form with long-lasting semi-double flowers. It does not set seed.

U. europaeus 'Strictus' is a slow-growing form which seldom flowers but makes an excellent low hedge.

PERENNIALS

Achillea millefolium 'Fanal'

Achilleas are hardy border plants which are very attractive to pollinators. They come in a range of colours and should be cut down to the base after flowering to avoid self-seeding. 'Fanal' has flat heads of red flowers in midsummer.

Mature height: 0.8m (2.5ft)
Mature spread: 0.4m (1.3ft)
Very hardy
Full sun

Achillea 'Terracotta' has apricot flowers fading to yellow.

Ulex europaeus.

Agapanthus 'Headbourne Hybrids'

These hybrids are of the deciduous group of *Agapanthus* which are the hardiest. They have lavender-blue flowers in rounded heads in summer and early autumn.

Mature height: 1m (3ft)
Mature spread: 0.5m (1.5ft)
Very hardy
Full sun

Agapanthus 'Loch Hope'

This is a hardy hybrid but in very cold areas may benefit from a mulch in winter. It has deep blue flowers in late summer.

Mature height: 1.5m (5ft)
Mature spread: 0.6m (2ft)
Hardy
Full sun

Achillea millefolium 'Fanal'.

Agapanthus praecox

This *Agapanthus* is evergreen in mild winters but deciduous in cold ones. It is better grown in containers in cold areas. It has sky-blue flowers in late summer.

Mature height: 0.8m (2.5ft)
Mature spread: 0.6m (2ft)
Hardy
Full sun

Anthemis punctata subsp. *cupaniana*

This *Anthemis* forms a low grey mat of foliage with white daisies all summer. It is best grown in full sun and well-drained soil, and is good for attracting pollinating insects.

Mature height: 0.3m (1ft)
Mature spread: 0.9m (3ft)
Very hardy
Full sun

Astelia chathamica 'Silver Shadow'

'Silver Shadow' is a useful silver-leaved plant for part shade and well-drained soils. It requires an acidic or neutral soil and is not a good choice for very cold areas.

Mature height: 1.5m (5ft)
Mature spread: 2m (6ft)
Hardy
Partial shade

Astelia nervosa 'Westland'

This is the hardiest *Astelia* with coppery bronze sheaves of leaves and small fragrant flowers.

Mature height: 0.6m (2ft)
Mature spread: 2m (6ft)
Hardy
Partial shade

Centranthus ruber

Red valerian is a clump-forming perennial which does best in poor soils and full sun. It is long flowering with red, white, or pink flower heads. It self-seeds freely and is likely to attract the hummingbird hawk-moth in southern areas of the United Kingdom.

Mature height: 1m (3ft)
Mature spread: 1m (3ft)
Fully hardy
Full sun

Crambe cordifolia

This is a clump-forming perennial which dies down in mid- or late summer. It can grow to at least 2m in height and likes a poor soil in a sunny, well-drained spot. It produces a vast froth of white scented flowers in late spring to early summer.

Mature height: 2.5m (8ft)
Mature spread: 1.5m (5ft)
Fully hardy
Full sun

Anthemis punctata subsp. *cupaniana*.

Astelia chathamica 'Silver Shadow'.

Crocosmia cultivars

Crocosmia are deciduous cormous perennials with sword-shaped leaves and branched spikes of funnel-shaped flowers in summer. They thrive in sun or part shade and preferably in a soil that retains some moisture.

Mature height: 0.75m (2.5ft)
Mature spread: 0.08m (0.25ft)
Hardy
Sun/partial shade

Crocosmia 'Limpopo' in effective combination with *Yucca* 'Golden Sword'.

C. 'Golden Fleece' forms spikes of lemon-yellow flowers in late summer.

C. 'Lady Hamilton' has apricot yellow flowers in late summer.

C. 'Limpopo' has large orange and pink flowers over a long period from August. In the garden they go well with agapanthus and make an excellent cut flower.

C. 'Lucifer' has red flowers on 1m high stems and is a highly recommended cultivar.

Centranthus ruber.

Crocosmia 'Lucifer'.

Erigeron glaucus

Beach aster forms dense mats of grey-green evergreen leaves and is covered from late spring to midsummer with pinkish-mauve flowers. It grows very well in coastal conditions but requires good drainage.

Mature height: 0.3m (1ft)
Mature spread: 0.45m (1.5ft)
Fully hardy
Full sun

Eryngium agavifolium

This is an unusual sea holly with spectacular rosettes of sharply toothed leaves. It likes a sunny, well-drained soil and grows to about 1.5m. Flowers in late summer with greenish flowers.

Mature height: 1.5m (5ft)
Mature spread: 0.6m (2ft)
Fully hardy
Full sun

Eryngium bourgatii 'Picos Blue'

This *Eryngium* is a more intense blue than the species. It has grey-green leaves with white veining.

Mature height: 1m (3ft)
Mature spread: 0.5m (1.5ft)
Fully hardy
Full sun

Erigeron glaucus.

Eryngium maritimum

This *Eryngium* is native in the United Kingdom. It likes sunny, well-drained soil and does well in shingle. It has spiny, leathery blue-green leaves and blue flowers.

Mature height: 0.9m (3ft)
Mature spread: 0.9m (3ft)
Fully hardy
Full sun

Eryngium × oliverianum

This hybrid *Eryngium* has large, spiky blue flowers on metallic blue stems from midsummer to early autumn. It does not transplant well, which is typical of *Eryngiums*.

Mature height: 0.9m (3ft)
Mature spread: 0.45m (1.5ft)
Fully hardy
Full sun

Eryngium agavifolium.

Eryngium × *tripartitum*

This is another hybrid *Eryngium* which flowers from midsummer to early autumn. The flowers are violet blue in colour.

Mature height: 0.9m (3ft)
Mature spread: 0.5m (1.5ft)
Fully hardy
Full sun

Eryngium zabelii 'Violetta'

This is a tap-rooted perennial which has blue flowers on blue stems in mid- and late summer.

Mature height: 0.70m (2ft)
Mature spread: 0.45m (1.5ft)
Fully hardy
Full sun

Kniphofia cultivars

Kniphofia are colourful, exotic looking plants with a long flowering season. They do well in coastal gardens in a well-drained, preferably sandy soil, in full sun or partial shade. However, they usually flower better in full sun.

Eryngium zabelii 'Violetta'.

Kniphofia 'Dorset Sentry'

This is an outstanding cultivar, one of the best known, has bright yellow flowers in summer and autumn.

Mature height: 1m (3ft)
Mature spread: 0.5m (1.5ft)
Fully hardy
Full sun/partial shade

Kniphofia 'Jenny Bloom'

This is a cultivar with pink flowers, which shade to cream, and are borne from late summer to mid-autumn.

Mature height: 1m (3ft)
Mature spread: 0.3m (1ft)
Fully hardy
Full sun/partial shade

Kniphofia 'Dorset Sentry'.

Kniphofia rooperi

This is a spectacular, late-flowering cultivar with huge, rounded orange and yellow pokers.

Mature height: 1.2 m (4ft)
Mature spread: 0.6 m (1ft)
Fully hardy
Full sun/partial shade

Kniphofia 'Royal Standard'

This cultivar has stout stemmed pokers of pale yellow flowers and light scarlet buds in late summer. It requires full sun in a well-drained soil. It may require some winter protection in very cold areas.

Mature height: 1m (3ft)
Mature spread: 0.6m (2ft)
Fully hardy
Full sun/partial shade

Kniphofia 'Tetbury Torch'

This cultivar has yellow to orange flowers in late summer and autumn and may have two flushes of flowers. It will grow in partial shade but flowers better in full sun.

Mature height: 1m (3ft)
Mature spread: 0.5m (1.5ft)
Fully hardy
Full sun/partial shade

Kniphofia 'Timothy'

'Timothy' is a cultivar which bears spikes of salmon-orange flowers from early summer to early autumn.

Mature height: 1m (3ft)
Mature spread: 0.5m (1.5ft)
Fully hardy
Full sun/partial shade

Kniphofia 'Wrexham Buttercup'

This cultivar has bright yellow flowers in midsummer.

Mature height: 1.2m (4ft)
Mature spread: 0.6m (2ft)
Fully hardy
Full sun/partial shade

Kniphofia rooperi.

Kniphofia 'Royal Standard'.

Osteospermum barberae (syn. *jucundum*)

This is a clump-forming perennial which likes a well-drained soil in full sun. It has a long flowering season with pink-purple daisies. It provides an excellent evergreen carpet.

Mature height: 0.3m (1ft)
Mature spread: 0.75 m (2.5ft)
Fully hardy
Full sun

Sisyrinchium striatum

This has iris-like leaves and cream flowers. It is best grown in a well-drained soil in full sun. Perennial and will propagate by self-seeding.

Mature height: 0.9m (3ft)
Mature spread: 0.3m (1ft)
Fully hardy
Full sun

Sisyrinchium striatum 'Aunt May'

This is similar to the species, but the leaves have cream margins. It is less vigorous than the species, will not set seed and can be propagated by division.

Mature height: 0.5m (1.5ft)
Mature spread: 0.3m (1ft)
Fully hardy
Full sun

Osteospermum barberae.

Sisyrinchium striatum.

Sisyrinchium striatum 'Aunt May'.

GRASSES

Grasses add a further dimension to the planting of the garden. They extend our range of architectural effects, and this, coupled with movement in the slightest breeze, makes them a very useful addition to many planting plans. A number are at their best in late summer and autumn.

Carex testacea

This sedge has leaves which are orange/bronzy yellow turning orange in full sun and greener in shade. It prefers some moisture in the soil but will not stand excessive winter wet. It is advisable to avoid very exposed positions.

Mature height: 1.5m (5ft)
Mature spread: 0.6m (2ft)
Hardy
Sun/partial shade

***Cortaderia* cultivars**

The pampas grasses form evergreen clumps which can be a little untidy and consequently may be best planted at the back of a deep border to hide most of the foliage. They prefer sunny, well drained soils and look best against dark backgrounds. It is best to protect the crowns of young plants in their first winter. There should be ample space left for them to develop.

Carex testacea.

Cortaderia fulvida

This is one of the showiest of the tussock grasses and likes some moisture in the soil. It is a tall species and forms one-sided plumes that are white or pink.

Mature height: 2.5m (8ft)
Mature spread: 2.5m (8ft)
Hardy
Full sun

Cortaderia richardii

This species is similar to *Cortaderia fulvida* and is hardy except in the coldest part of the United Kingdom. It has one-sided plumes that are creamy turning white, which start to form in early to midsummer and persist until the winter. Many people consider it to be the best pampas grass.

Mature height: 2.5m (9ft)
Mature spread: 1.8m (6ft)
Hardy
Full sun

Cortaderia selloana

This pampas grass produces pyramidal to oblong panicles of flowers in late summer. The plumes are often flushed with pink or purple.

Mature height: 2.5m (9ft)
Mature spread: 1.5m (5ft)
Very hardy
Full sun

Cortaderia selloana 'Pumila'

This excellent cultivar, suitable for smaller gardens, is probably the finest pampas grass when in flower. The plumes are a silvery-yellow colour.

Mature height: 1.5m (5ft)
Mature spread: 1.2m (4ft)
Very hardy
Full sun

Cortaderia selloana 'Sunningdale Silver'

This cultivar has silvery-white plumes on stout stems which are sturdy and wind resistant.

Mature height: 3m (10ft)
Mature spread: 2.5m (8ft)
Very hardy
Full sun

Leymus arenarius

This grass is a runner (invasive) so plant appropriately. However, it was Gertrude Jekyll's favourite grass. It is virtually evergreen in warmer areas. It likes to be in full sun and has very blue foliage.

Mature height: 1.5m (5ft)
Mature spread: indefinite.
Fully hardy
Full sun

Cortaderia selloana 'Pumila'.

Leymus arenarius.

Panicum virgatum

The switch grasses are deciduous and form upright clumps reaching 0.9–2m (3–6.5 feet) depending on the cultivar. Blue-leaved cultivars require perfect drainage and full sun. Green cultivars are less fussy. The plants are best left until the end of winter before they are cut down as they can provide a good winter feature when covered in hoar frost.

Mature height: 1.2m (4ft)
Mature spread: 0.75m (2.5ft)
Fully hardy
Full sun/partial shade

P. virgatum 'Heavy Metal' has glaucous blue stems and pink plumes.

P. virgatum 'Prairie Sky' is the bluest of all and has amber-brown flowers.

P. virgatum 'Shenandoah' is unmatched for its burgundy leaf colour in late summer. It reaches 1m high.

Panicum virgatum.

BRITISH NATIVE PLANTS

These plants are to be found growing along the edge of the coast in the United Kingdom. They are worth consideration for planting in shingle gardens and by those gardeners who wish to make their garden attractive to wildlife. Many of these native, coastal British plants are important food sources for the caterpillars of a range of moths, and we have listed some of these below.

Althaea officinalis

Marsh mallow likes moist soils. It has pinkish flowers in late summer which are extremely attractive to various bees.

- Food plant: marsh mallow moth and pale shoulder moth.

Armeria maritima

Thrift thrives in poor soil in full sun and is commonly found growing wild along the British coast. It forms a low mound of evergreen foliage and has small pink flowers from spring to summer.

A. maritima 'Alba' is a white form of the plant.

- Food plant: thrift clearwing and black-banded wave moths.

Artemisia vulgaris

Mugwort has grey foliage which provides protection from salt and wind. It grows to about 1.5m tall and has white flowers in summer.

- Food plant: mullein wave, wormwood pug and bordered pug moths.

Artemisia 'Oriental Limelight' is an attractive variegated form but is invasive so plant carefully.

Aster tripolium

Sea aster is a perennial with small purple and yellow flowers in late summer. It is often visited by bees and flies.

- Food plant: rosy wave and star wort moths.

Armeria maritima.

Beta vulgaris subsp. *maritima*

Sea beet is an annual or short-lived perennial with small, insignificant flowers in summer. It is edible and of interest to foragers, with large fleshy leaves which can be used like chard.

Centranthus ruber

Red valerian likes hot dry soils and has a long flowering season. It has red, pink, or white flowers which the humming-bird hawk-moth and silver Y moth find very attractive.

- Food plant: large ranunculus moth.

Crambe maritima

Sea kale, a fantastic plant typical of shingle beaches, requires a sunny, very well drained spot. It can be blanched for food but does have spectacular flower heads. It dies down for the winter.

Cynoglossum officinale

Hound's tongue is a biennial with red-purple flowers in summer.

- Food plant: scarlet tiger moth.

Daucus carota

Wild carrot is a biennial on sunny sites and chalky soils. It is often found by the sea. It flowers from June to August.

- Food plant: yellow belle moth.

Eryngium maritimum

Sea holly requires a sandy, well-drained soil in full sun. It does not like being moved. It has glaucous foliage and pinkish blue flower heads.

Beta vulgaris subsp. *maritima*.

Crambe maritima.

Centranthus ruber.

Daucus carota.

Foeniculum vulgare

Fennel, widely used as a culinary herb, can also be grown for its feathery foliage. It is a prolific self-seeder so do not allow it to set seed. It can be cut down to the base immediately the yellow flowers fade and it will have fresh foliage by the end of summer. It is a good nectar plant.

F. vulgare 'Purpureum' is an attractive cultivar with purple tinged foliage.

Galium verum

Lady's bedstraw is a low-growing perennial that requires a sunny site and is usually found in meadows. It has yellow flowers in July and August.

- Food plant: riband wave, oblique striped and bedstraw hawk moths.

Foeniculum vulgare.

Foeniculum vulgare 'Purpureum'.

Glaucium flavum

The yellow horned poppy is not long lived but will self-seed. It grows well in shingle. It has yellow flowers throughout the summer followed by long, curved seed pods. It does not respond well to being transplanted.

Jasione montana

Sheep's bit scabious is an annual or short-lived perennial for well-drained, lime-free soils. It is a good bee plant.

- Food plant: coast dart moth.

Koeleria macrantha

Crested hair grass is found mostly on chalk or limestone soils or sand dunes. In winter, the leaves can turn quite blue.

- Food plant: most moths that use grasses as larval food plants are likely to use this species.

Lavatera maritima

The tree mallow is a semi-evergreen shrub with grey leaves and pink flowers. It is hardy in coastal areas but is a short-lived plant.

Leymus arenarius

Refer to section on grasses for details.

- Food plant: lyme grass moth.

Glaucium flavum.

Limonium binervosum

Rock sea lavender is a low-growing perennial for a sunny, well-drained position. It does well in a shingle garden. It has lavender flowers in late summer.

- Food plant: ground lackey moth.

Limonium vulgare

Sea lavender is a perennial with lavender flowers in late summer.

Lotus corniculatus

Bird's foot trefoil is a spreading perennial with yellow flowers. It is found in meadows and will require some protection from coastal exposure.

- Food plant: six-belted clearwing and six-spot burnet moths.

Oenothera biennis

Evening primrose is a tall, night-flowering biennial with yellow flowers in summer. It prefers a position in full sun and a well-drained soil. It is pollinated by moths.

- Food plant: willow hawk moth. This moth is an extremely rare visitor to the United Kingdom.

Ononis repens

Restharrow is a low-growing perennial which grows well in chalky soils. It has pink flowers in summer.

- Food plant: rest harrow moth.

Plantago coronopus

Buck's horn plantain is a native of coastal areas. It has brownish flowers from April to July.

- Food plant: yellow belle moth.

Plantago lanceolata

Although the ribwort plantain is not the most attractive wild-flower, it is a good plant for attracting moths and is the food plant for several species. It is also an effective treatment for nettle stings.

Plantago maritima

Sea plantain has greenish yellow flowers from June to August. It is a native of salt marshes.

- Food plant: ground lackey and mullein wave moths.

Rhodiola rosea

Roseroot is a perennial with grey leaves and yellow flowers. It is extremely attractive to bees.

Saponaria officinalis

Soapwort is an attractive perennial with pink flowers in late summer. It can be very invasive.

- Food plant: campion moth.

Oenothera biennis.

Plantago lanceolata growing on a clifftop amongst thrift (*Armeria maritima*).

Sedum album.

Sedum acre

Biting stonecrop is a low-growing perennial with yellow flowers in spring and early summer. It performs well grown in gaps in paving.

- Food plant: feathered ranunculus moth.

Sedum album

White stonecrop is a low-growing perennial for dry, sunny soils. It is a good bee plant. In the United Kingdom it is most common in the south-west.

- Food plant: Scotch annulet moth.

Sedum anglicum

English stonecrop is a low-growing perennial with pink flowers in summer. It does well in shingle.

- Food plant: Apollo and yellow-ringed carpet moths.

Silene maritima.

Silene maritima

Sea campion requires a well-drained soil in a sunny spot to perform well. It forms a low mound of bluish-green foliage with attractive white flowers and works well on a rockery.

- Food plant: marbled coronet and netted pug moths.

Thymus serpyllum

Creeping thyme requires a sunny, well-drained soil. It forms a very low mat of evergreen foliage, excellent for planting in gaps in paving, and its purplish or white flowers are extremely attractive to bees and butterflies.

Thymus serpyllum.

SUCCULENTS

These plants are for gardeners who do not mind taking a risk. They are only for consideration by those who live in milder climates as they will not withstand prolonged spells of frost.

Agave americana

This is an evergreen and requires full sun and sharp drainage. It has broad, glaucous leaves with sharp spiny edges. It does not grow well on chalky soils. Established specimens will look gnarled and scarred but that is part of their attraction. It can be propagated by removing rooted offsets. It can make for a major feature, or focal point, in the coastal garden.

Mature height: 2m (6ft)
Mature spread: 3m (10ft)
Hardy in mild areas
Full sun

A. americana 'Marginata' has an outer band of creamy yellow variegation.

Beschorneria septentrionalis

This needs to be in full sun in a well-drained spot with some shelter. *Beschorneria* cultivars are only for gardens with a mild climate as they will not tolerate prolonged frost. It has a rosette of green leaves and dramatic flower panicles in late spring or summer.

Mature height: 1.5m (5ft)
Mature spread: 1.5m (5ft)
Hardy in mild areas
Full sun

Beschorneria yuccoides 'Quicksilver'

This *Beschorneria* requires the same conditions as *B. septentrionalis* but is a more compact plant with a rosette of silver-green leaves. In summer it bears panicles of yellow-green flowers with red bracts.

Mature height: 1.2m (4ft)
Mature spread: 1m (3ft)
Hardy in mild areas
Full sun

Fascicularia bicolor

This is an evergreen perennial requiring full sun and sharp drainage in poor soils. It may require some protection from wet weather in winter. Mature rosettes have dense corymbs of pale blue flowers with bright crimson inner leaves.

Mature height: 0.45m (1.5ft)
Mature spread: 0.6m (2ft)
Hardy in mild areas
Full sun

Agave americana.

Fascicularia bicolor.

Yucca filamentosa

Yucca filamentosa requires a well-drained soil in full sun, as do all yuccas. It has a clump-forming habit and leaves with curly white threads. In mid- to late summer, it produces upright panicles of white flowers which can reach up to 2m in height.

Mature height: 0.75m (2.5ft)
Mature spread: 1.5m (5ft)
Very hardy
Full sun

Y. filamentosa 'Bright Edge' is similar to the above but with broad yellow leaf margins.

Yucca filamentosa 'Bright Edge'.

Yucca flaccida

This is a clump-forming, almost stemless shrub. It bears panicles of white flowers in mid- and late summer. The panicles can reach about 1.5m high.

Mature height: 0.5m (2ft)
Mature spread: 1.5m (5ft)
Very hardy
Full sun

Y. flaccida 'Golden Sword' forms a stemless clump of green leaves with a central gold band.

Y. flaccida 'Ivory Tower' is stemless with grey-green leaves and large white panicles of flowers.

Yucca gloriosa

This is an erect shrub with tall panicles of flowers from late summer to autumn.

Mature height: 2m (6ft)
Mature spread: 2m (6ft)
Hardy
Full sun

Y. gloriosa 'Variegata' has a stemmed habit with spiky cream and pink striped foliage.

Yucca flaccida 'Golden Sword' in very effective combination with *Agave americana*.

ANNUALS

Hybrid Gazania

Hybrid gazanias are strictly perennials but are not likely to survive the winter outdoors, so, in the United Kingdom, they are usually treated as annual bedding plants. However, they may self-seed in a protected spot. They come in a good range of colour and can be used as bedding plants to good effect e.g., orange gazania plus blue agapanthus.

Gazania hybrids.

Mesembryanthemum criniflorum* syn. *Dorotheanthus bellidiformis

The Livingstone daisy is a low-growing annual which comes in a wide range of colours. It grows 10–15cm (0.5 feet) high and requires a well-drained soil, with low fertility in full sun. Seed is best sown under cover in spring or buy young plants for planting out in May.

HARDY ANNUALS

In spaces where plants have been left to develop, seed of hardy annuals can be sown to produce patches of colour until the main planting has matured and filled in. They can be left to self-seed at the end of the summer where they will find niches that suit them. If too many seedlings are produced, they can easily be uprooted whilst small. The overall effect will be an informal one. Examples that should be successful include:

Calendula 'Snow Princess'.

***Calendula officinalis* (pot marigold)**

Seed can be sown in the autumn, otherwise in the spring when it will flower in about ten weeks from sowing. The species has bright orange single flowers but there are a number of very attractive named cultivars.

***Centaurea cyanus* (cornflower)**

Seed can be sown in the autumn or spring. It should be sown in the spaces where it is to flower. It will flower all summer if dead headed. A range of colours is available.

Cosmos sulphureus with *Agapanthus*.

***Cosmos sulphureus* (yellow cosmos)**

This is a half-hardy annual but is included in this section as it is well worth trying in a coastal setting. Seed should be sown in mid-spring. The flowers are usually yellow or orange. It is important to dead head the plants in order to keep them flowering throughout the summer.

***Eschscholzia californica* (Californian poppy)**

Seed is best sown in the autumn, for early flowering the following year. They should be sown in their intended flowering position, rather than in seed trays for transplanting. The poorer the soil the better it flowers. Although the flowers are usually orange there is a range of colours available. The plant usually self-seeds.

***Papaver rhoeas* (field poppy)**

Seed is best sown in the autumn where it is to flower. Dead heading is essential to prolong flowering. The Shirley poppy was developed from the wild species and seed is available in a good range of colour. Just sow the seed and forget them!

***Papaver somniferum* (opium poppy)**

This is best grown in full sun and reaches about 1m (3 feet) tall, producing large flowers. Seed of a range of colours, and of single- and double-flowered varieties, is available. They should be sown in autumn or spring and dead headed to prolong flowering.

Eschscholzia californica.

CLIMBERS

Climbing plants are not great in coastal conditions. They have not evolved to become resistant to strong coastal, salt winds. The climbers suggested below may be successful if grown in a sheltered position.

Campsis radicans

The trumpet vine is hardy but requires some shelter from strong winds. It can be pruned hard in early spring before growth starts. It is a deciduous climber and has orange-red flowers in summer and autumn.

Mature height: 10m (30ft)
Hardy
Full sun

***Campsis × tagliabuana* 'Madame Galen'**

This is a vigorous climber and bears panicles of orange-red trumpet-shaped flowers in late summer and autumn.

Mature height: 10m (30 ft)
Hardy
Full sun

Eccremocarpus scaber

The Chilean glory vine is often grown as an annual but is perennial. It likes a well-drained soil in full sun. Seed can be sown in early spring under glass, and it should be planted out in May when all danger of frost has passed. It bears orange-red flowers from late spring to autumn and remains evergreen in coastal gardens.

Mature height: 5m (15ft)
Hardy
Full sun

Hydrangea petiolaris

This hydrangea is a vigorous deciduous climber which clings to walls and fences by aerial roots. It has white flowers in summer and requires little pruning but can be trimmed to fit the available space.

Mature height: 15m (50ft)
Fully hardy
Full sun/partial shade

Lonicera henryi

This is a vigorous evergreen honeysuckle which should be grown in good soil in partial shade. It has yellow-throated purplish-red flowers in early to midsummer, and these are followed by black berries.

Mature height: 10m (30ft)
Very hardy
Partial shade

Muehlenbeckia complexa

This is a vigorous, deciduous climber with small dark green leaves. It has greenish-white flowers in summer which are followed by fleshy white fruit.

Mature height: 3m (10ft)
Hardy
Full sun/partial shade

GREAT PLANT COMBINATIONS

In this section are listed some good combinations of plants whose colour, form and habit go well together. You may find them to be a useful starting point for your planting design. The following section provides some more detailed planting schemes in a variety of styles.

Chalk soils

Chalk soils are alkaline and therefore ericaceous plants cannot be considered. If the soil is shallow, then it is likely to be free draining and low in nutrients. The quality of the soil can be improved by the addition of plenty of homemade compost or soil improver.

Combination 1
Olearia × scilloniensis
Cistus × argentatus 'Silver Pink'
Fuchsia magellanica 'Hawkshead'
Genista hispanica

Combination 2
Hebe × franciscana 'Blue Gem'
Rosa rugosa 'Alba'
Lavandula angustifolia 'Hidcote'
Artemisia 'Powis Castle'

Combination 3
Cortaderia selloana 'Pumila'
Yucca flaccida 'Ivory Tower'
Crocosmia 'Lucifer'
Centranthus ruber

Clay soils

Although clay soils can be difficult to work, they are very fertile. For details concerning gardening on clay soils, refer to the soil preparation section in Chapter 4. There is a wide range of plants that are perfectly happy growing on clay soil.

Combination 1
Tamarix tetrandra
Cistus purpureus
Rosa rugosa
Berberis × ottawensis

Combination 2
Rosa rugosa 'Fru Dagmar Hastrup'
Kniphofia rooperi
Panicum virgatum 'Shenandoah'
Crocosmia 'Emberglow'

Combination 3
Escallonia 'Apple Blossom'
Rosa 'Stanwell Perpetual'
Hydrangea macrophylla 'Sea Foam'
Cistus × purpureus 'Alan Fradd'

Yellow and gold

The use of yellow foliage can brighten up a planting scheme and provide a useful contrast. Very open situations should be avoided if sun scorch is a possible problem, as is the case with *Berberis thunbergii* 'Aurea'. However, too much shade may result in lime green foliage rather than yellow. Addition of some purple foliage will increase the interest. Some gardeners may find the mix of yellow foliage and deep pink of the flowers of the *Escallonia* a bit of a clash, but it is certainly bold and eye catching.

Griselinia littoralis 'Variegata'
Escallonia laevis 'Gold Ellen'
Berberis thunbergii 'Aurea'
Foeniculum vulgare 'Purpureum'

Purple and plum

Purple works well with orange and gold and can be very complementary to bright yellow. Purple and yellow as different foliage types or at different heights in the planting can be highly effective. Addition of strong purples to a palette of pastel colours can add depth and effective contrast. Red foliaged plants such as *Berberis*, used carefully, can work harmoniously with the oranges of a hot colour scheme.

Combination 1
Phormium tenax 'Purpureum'
Buddleja globosa
Yucca flaccida 'Golden Sword'
Crocosmia 'Emberglow'

Combination 2
Phormium 'Platt's Black'
Berberis thunbergii 'Atropurpurea Nana'
Lavandula angustifolia 'Hidcote'
Santolina chamaecyparissus

Combination 3
Sambucus nigra 'Black Beauty'
Berberis × ottawensis
Artemisia 'Powis Castle'
Lavandula angustifolia 'Hidcote'

Silver and gold

Many silver-leaved plants are found in the Mediterranean region. They thrive in full sun and are drought resistant. The soil should be well drained as they dislike standing in wet soils during the winter.

Combination 1
Elaeagnus 'Quicksilver'
Agapanthus 'Headbourne Hybrids'
Panicum virgatum 'Heavy Metal'
Eschscholzia californica

Combination 2
Cistus × argentatus 'Silver Pink'
Convolvulus cneorum
Genista hispanica
Rosmarinus officinalis 'Prostratus'

Hot and dry

As a result of global warming, drought-resistant plants are becoming increasingly important when considering the planting in our gardens. This is even more important in coastal gardens, with windy conditions producing increased transpiration. It is important that any new planting, even of drought-resistant plants, is thoroughly watered until the plants are well established.

Combination 1
Spartium junceum
Cortaderia richardii
Cistus × purpureus 'Alan Fradd'
Phlomis fruticosa

Combination 2
Cistus purpureus 'Alan Fradd'
Convolvulus cneorum
Eryngium oliverianum
Lupinus arboreus

Combination 3
Genista hispanica
Cistus × aguilarii 'Maculatus'
Phlomis fruticosa
Rosmarinus officinalis 'Miss Jessopp's Upright'

Native plants

Several attractive native plants thrive in coastal areas and will help to increase the biodiversity within the garden.

Combination 1
Armeria maritima
Glaucium flavum
Silene maritima
Eryngium maritimum

Combination 2
Foeniculum vulgare
Centranthus ruber
Saponaria officinalis
Oenothera biennis

Exotics

This combination is for gardeners who are prepared to take a risk with their planting. They need to be planted in areas with a mild climate which do not experience prolonged spells of frost.

Chamaerops humilis
Agave americana
Yucca filamentosa
Agapanthus praecox

SAMPLE PLANTING SCHEMES

Native plants

This bed will require a sunny, well-drained spot. Horticultural grit could be added to improve the drainage if necessary. A top dressing of shingle after planting will create a naturalistic effect. The planting area will be attractive during spring and summer. As these are plants which grow in the wild, they will not require the addition of any fertiliser. Some of the plants may be difficult to source but there are nurseries which specialise in the provision of plugs of native plants. The plugs establish quickly but will require watering in dry spells whilst they do so. In larger areas these can be planted at a rate of seven per square metre.

Rosa 'Stanwell Perpetual' is a horticultural form of the native burnet rose (*Rosa pimpinellifolia*); the tree lupin (*Lupinus arboreus*) is not a native plant but is naturalised along much of the south coast of England.

Sea kale (*Crambe maritima*) is a spectacular plant when in flower, and fennel (*Foeniculum vulgare*) adds height to the planting.

The sea kale and the sea holly (*Eryngium maritimum*) have long tap roots and will not move well so their position in the bed requires careful thought.

Common names for the other native plants include yellow horned poppy (*Glaucium flavum*), rock sea lavender (*Limonium binervosum*), Breckland thyme (*Thymus serpyllum*), thrift (*Armeria maritima*), and ribwort plantain (*Plantago lanceolata*). The latter, although not particularly attractive when in flower, is the food plant for several moth caterpillars and is worth growing from a wildlife point of view. This is also true for several of the other plants, details of which can be found in the plant section on plants native to the United Kingdom. It is good practice to plant several specimens of each plant as they are more likely to be found by insects and it makes the appearance of the planting more natural.

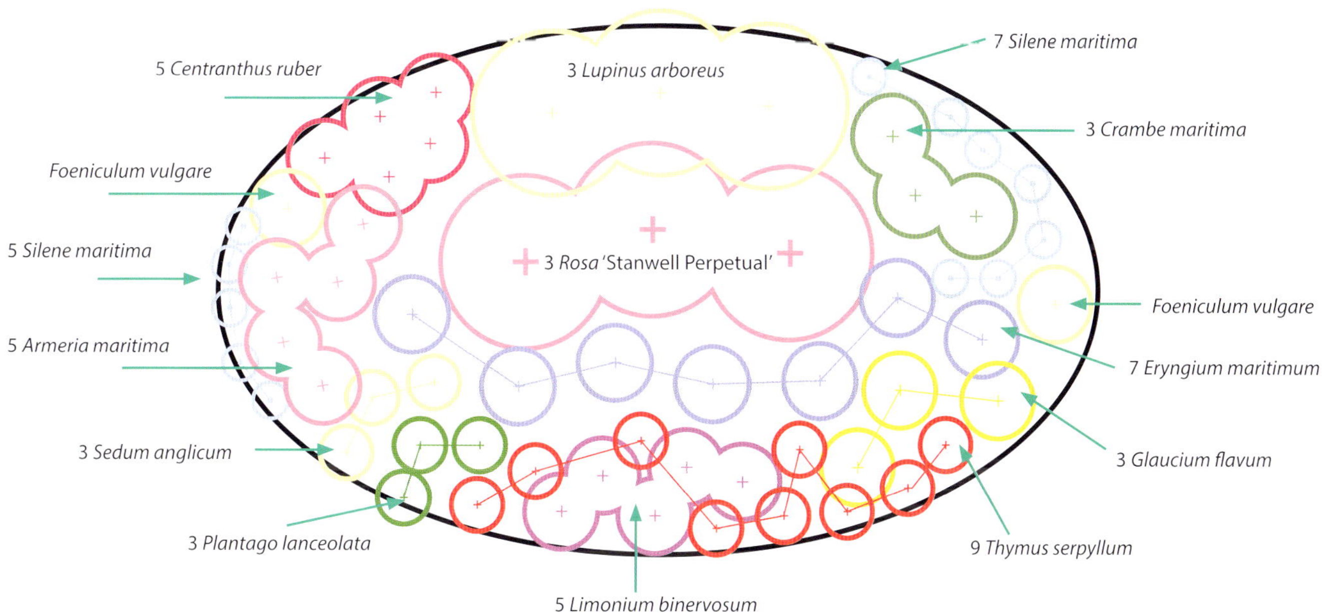

Native plants sample planting scheme.

Exotics

This planting is for gardeners who like to grow the more unusual plants. Several of the plants featured here are only for consideration by those who live in areas with milder climates as they will not withstand prolonged spells of frost. The soil should be well drained and in full sun. Drainage can be improved by the addition of large quantities of horticultural grit.

This scheme would work well as a gravel garden and rocks could be added to give the effect of a shingle beach. There are more details of how to produce this effect in Chapter 3 (hard landscaping).

Agave americana has very sharp, spiny edges and this needs to be kept in mind if young children are likely to encounter them. As it ages, it will look gnarled and scarred but this is part of its attraction. The spines on the yuccas can be carefully removed with a pair of nail clippers without spoiling the appearance of the plant, if this a concern. *Astelia chathamica* has attractive silver foliage, the appearance of which is enhanced by the dwarf *Phormium* 'Bronze Baby'. *Chamaerops humilis* is considered by many to be the best palm for coastal areas. It does not become too large and will look best if not crowded by other planting.

Silver and gold

The plants that feature in this design require a well-drained soil in full sun. The planting is of shrubs and perennials, including grasses, and will look its best during summer and early autumn. *Cortaderia selloana* 'Sunningdale Silver', an important feature of the design, is an impressive grass which grows up to 3m high. The plumes of flower are wind resistant.

Brachyglottis 'Sunshine' has good silvery foliage and a long flowering season. It can be pruned at the end of the winter to keep it compact. Other silver foliage plants in the scheme include *Santolina chamaecyparissus* 'Nana' and *Convolvulus cneorum*. The *Santolina* is included for the effect of its foliage. It has yellow flowers, but the suggestion is to trim these off with a pair of shears as the buds start to open. This will stop the structure of the plant from opening up and will produce a compact mound of attractive silver foliage. *Convolvulus cneorum* also has attractive silver foliage and a very long flowering season. The white flowers will be a very attractive feature. The foliage is likely to look a bit untidy at the end of the winter and the plant should be tidied up with a pair of secateurs.

Rosa rugosa 'Alba' is a tough, disease-free rose. It should not require any spraying with insecticides or fungicides and indeed may well react badly to the application of such sprays. Nor does it require dead heading of flowers as it will form red hips which adds to the attraction in late summer and autumn.

Panicum virgatum 'Northwind' has fine, blue-green foliage. The old flower stems should not be removed until the end of the winter as they are an attractive feature when covered in hoar frost. When the old stems are removed, they can be cut back hard but take care not to damage the new fresh ones as they emerge from the soil.

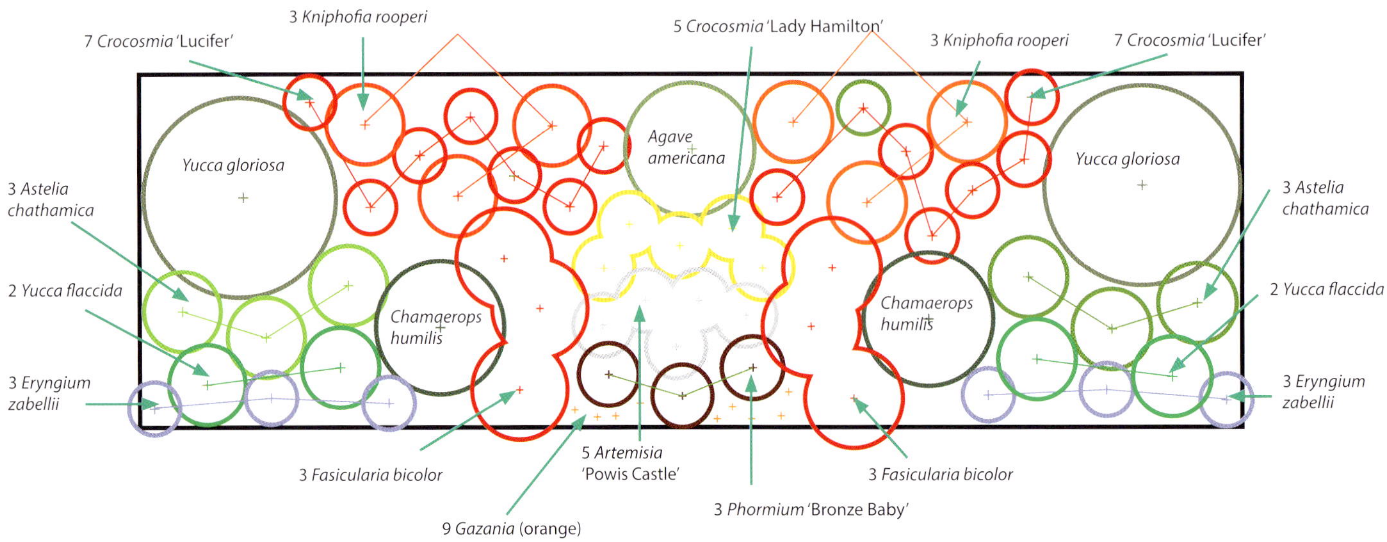

Exotic plants sample planting scheme.

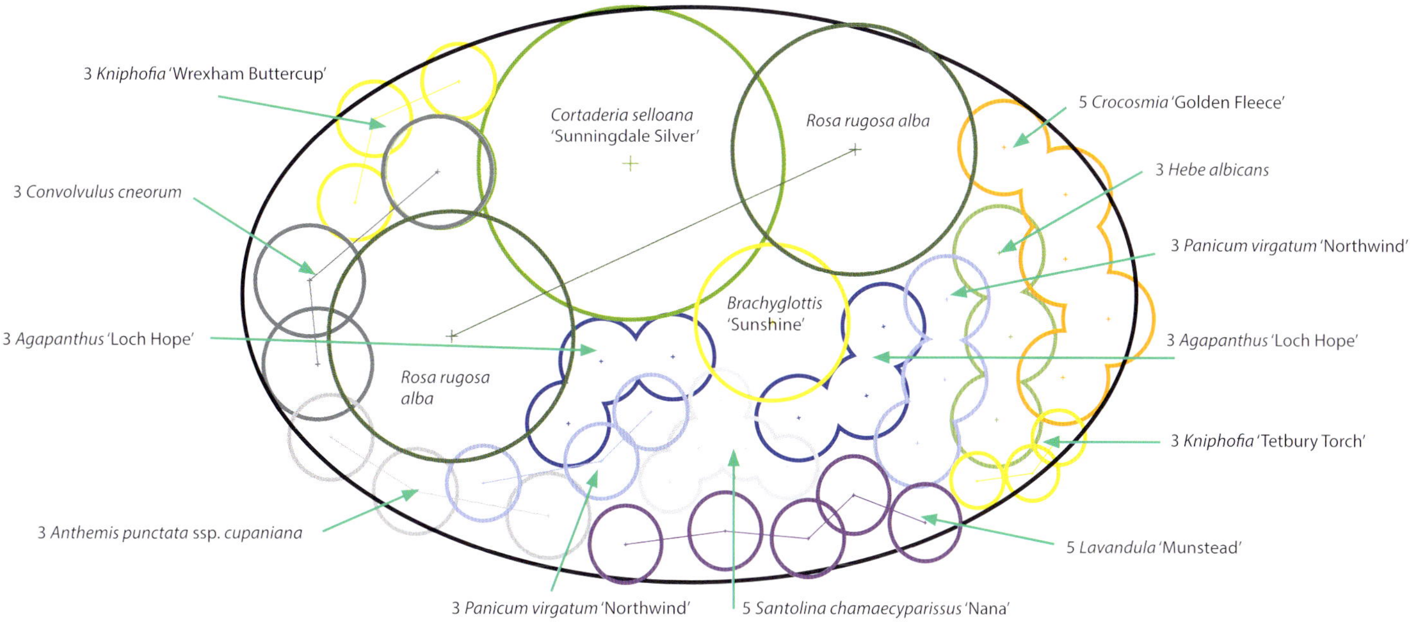

Silver and gold sample planting scheme.

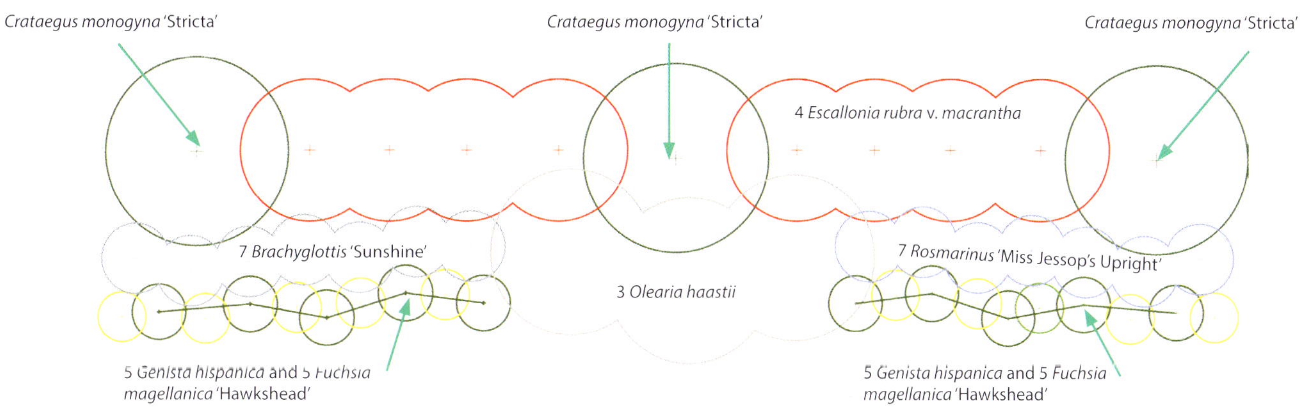

A small windbreak sample planting scheme.

Further accents of blue are added by the agapanthus and lavender. *Lavandula* 'Munstead' will require pruning, but not too hard, either after flowering or at the end of the winter.

Small windbreak

The rear of this planting scheme contains the tallest plants and should face into the prevailing winds. It is important that the plants are fed with a slow-release fertiliser, such as fish, blood and bone, and watered well during the first season. The size of the plants decreases as we move forward across the bed and the overall effect should be that of a wide bed that effectively filters the wind.

Crataegus monogyna is one of few trees suitable for the small garden that can cope well in a coastal garden. *Crataegus monogyna* 'Stricta' is a fastigiate form, restricted in width with a columnar form. It is very suitable for growing as an architectural accent in a hedge, which in this example is composed of *Escallonia rubra* v. *macrantha*. This evergreen forms a good hedge and has red flowers. It can be trimmed as required but can grow to a height of about 4m.

Olearia haastii is one of the finest evergreens for coastal regions. It has white flowers which will show up well against the glossy foliage of the *Escallonia*. The remaining shrubs should be encouraged to become bushy by suitable pruning and add good summer colour to the planting. They can all be pruned, if necessary, after flowering or at the end of the winter.

Large windbreak

This scheme utilises the same ideas as the smaller version but on a larger scale. The tallest plants are at the rear, facing the prevailing wind.

Cupressus macrocarpa is eventually a large tree and only suitable for large gardens. It can, however, make a decent hedge if kept cut. The cultivar 'Goldcrest' is far denser and more compact, but it can scorch in very exposed sites.

Elaeagnus ebbingei is completely hardy and fast growing and makes an excellent hedge with small, scented flowers in autumn. The cultivar 'Gilt Edge' has the leaves edged golden yellow.

The flowers of *Atriplex halimus* are insignificant but it has good silvery foliage some of which it may lose during the winter in very cold areas. It makes a good repeating feature here, contrasting well with the yellow flowers and purple foliage of the *Berberis × ottawensis*. If the *Atriplex* has retained its foliage during the winter, the contrast with the foliage of the *Berberis* will be retained as a good winter feature.

Brachyglottis 'Sunshine' also has good silvery foliage to add further contrast with the *Berberis*. It also has a long flowering season with yellow flowers which go well with the violet-blue flowers of the *Hebe franciscana* 'Blue Gem' and the blue flowers of the tree lupin (*Lupinus arboreus* 'Blue Boy'). *Brachyglottis* 'Sunshine' should be pruned annually to keep it reasonably compact. The tree lupin is not particularly long lived but is likely to self-seed or can be replaced when necessary.

Rosa 'Stanwell Perpetual' blooms all summer with good fragrance and will require an annual prune but not too hard a one. It is being grown as a medium-sized shrub.

If room permits, *Ulex europaeus* 'Flore Pleno' could be added to the front row of the planting. This is an outstanding cultivar of the native gorse with a long flowering season of yellow, semi-double flowers; it does not self-seed.

The use of several repeating units of a limited palette of plants provides a visually impressive and colourful windbreak.

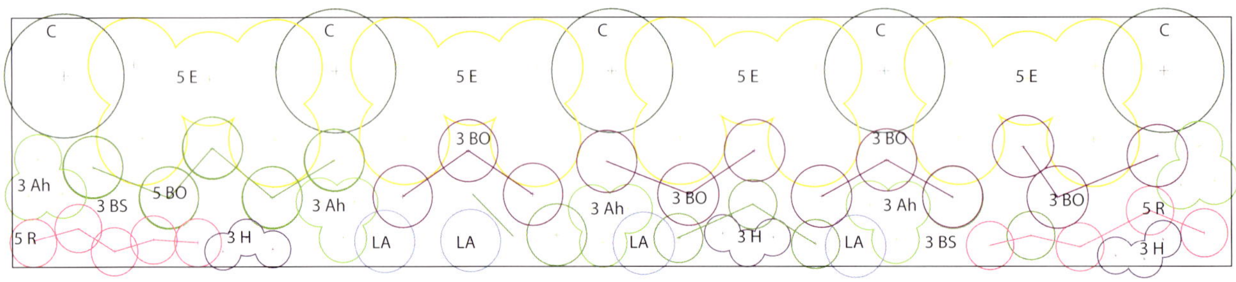

Ah – *Atriplex halimus*
BO – *Berberis × ottawensis*
BS – *Brachyglottis* 'Sunshine'
C – *Cupressus macrocarpa*
E – *Elaeagnus × ebbingei* 'Gilt Edge'
H – *Hebe franciscana* 'Blue Gem'
LA – *Lupinus arboreus* 'Blue Boy'
R – *Rosa* 'Stanwell Perpetual'

A large windbreak sample planting scheme.

Mediterranean

The plants featured in this design are mainly drought-tolerant and like a well-drained soil in full sun. Three pairs of *Cistus* form the structural backbone. They are excellent evergreens and fully drought-tolerant but do not respond well to pruning.

Between the pairs of *Cistus* are the larger plants which will give some shelter to those in front if they are planted to the windward side of the bed. *Cortaderia richardii* reaches over 2 metres in height. *Spartium junceum* grows to about 3 metres. The latter has masses of yellow flowers in summer. It can become a little untidy so may require a light prune in early spring.

Genista hispanica also has yellow flowers and is deciduous.

The neighbouring planting of *Panicum virgatum* flowers in late summer and can provide an attractive winter feature when covered in frost. It requires cutting back to the base at the end of winter taking care to avoid damaging any new growth. More yellow is added by *Phlomis fruticosa*, an evergreen with a long flowering season.

The blue of the *Panicum* foliage is picked up by agapanthus, rosemary and lavender. *Lavandula* 'Hidcote' can be burnt by strong salt winds so will require some shelter in very exposed situations. *Convolvulus cneorum* is an excellent companion to it and has a long flowering season and attractive silver foliage. The theme of silver foliage is continued with *Artemisia* 'Powis Castle' and *Santolina chamaecyparissus*, which should be grown as foliage plants. This can be achieved by pruning the *Artemisia* quite hard at the end of the winter and by not allowing the *Santolina* to flower by shearing off the flower buds before they open.

The *Erigeron glaucus* and *Osteospermum barberae* have long flowering seasons and will produce carpets of colour. The use of natural rock completes the Mediterranean coastline theme.

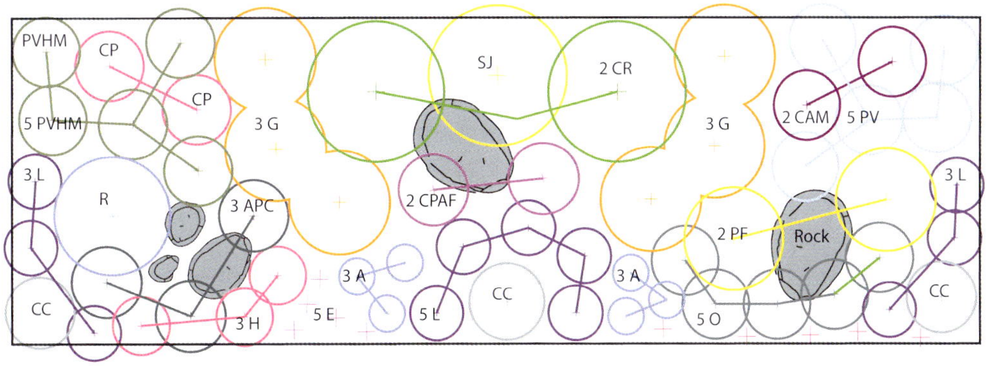

A – *Agapanthus* 'Headbourne Hybrids'
APC – *Artemisia* 'Powis Castle'
CAM – *Cistus aguilarii* 'Maculatus'
CP – *Cistus* × *purpureus*
CPAF – *Cistus* × *purpureus* 'Alan Fradd'
CC – *Convolvulus cneorum*
CR – *Cortaderia richardii*
E – *Erigeron glaucus*
G – *Genista hispanica*
H – *Hebe* 'Wiri Image'
L – *Lavandula* 'Hidcote'
O – *Osteospermum barberae*
PF – *Phlomis fruticosa*
PV – *Panicum virgatum* 'Prairie Sky'
PVHM – *Panicum virgatum* 'Heavy Metal'
R – *Rosmarinus officinalis* 'Miss Jessop's Upright'
SC – *Santolina chamaecyparissus*
SJ – *Spartium juncium*

Mediterranean plants sample planting scheme.

Pastel

The windward side of the bed, which should be a sunny, well-drained soil, contains the taller plants which will filter the wind and provide some shelter to the plants in front, particularly the lavenders. *Lavandula* cultivars such as 'Munstead' can be scorched, or even killed, in very exposed sites.

Escallonia 'Apple Blossom' has evergreen leaves and pink and white flowers in summer.

The *Tamarix tetrandra* has light pink flowers which are on the previous year's growth. An alternative is *Tamarix ramosissima* 'Pink Cascade' which flowers a little later. They are pruned at different times; *see* plant list for details.

The pink theme continues with *Fuchsia magellanica* var. *molinae* which has very pale pink flowers. If it becomes too wide, it can be pruned to the base at the end of winter.

The corners of the scheme feature cultivars of *Hydrangea macrophylla*. *Hydrangea* 'La Marne' is a mophead hydrangea with large blue flowers if in acidic soil, otherwise they will be pink. *Hydrangea* 'Sea Foam' is a lacecap hydrangea, also with blue or pink flowers. Both of these hydrangeas are accompanied by the North American switch grass, *Panicum virgatum* 'Heavy Metal' which also has pink plumes of flower. It is at its best in late summer and autumn when the *Kniphofias* should both be adding to the overall effect.

The central planting has *Cortaderia selloana* 'Pumila' on the windward edge. This is considered by many to be the finest pampas grass when in flower. It will form a good backdrop to the planting of *Agapanthus praecox* and the cream-flowered *Sisyrinchium striatum* 'Aunt May'. The *Sisyrinchium* contrasts well with the lavender-blue flowers of *Lavandula* 'Munstead' which are complemented by the silver foliage of *Convolvulus cneorum*. The two plantings of this combination are separated by the blue *Eryngium bourgatii* 'Picos Blue', which adds further contrast to the *Sisyrinchium*.

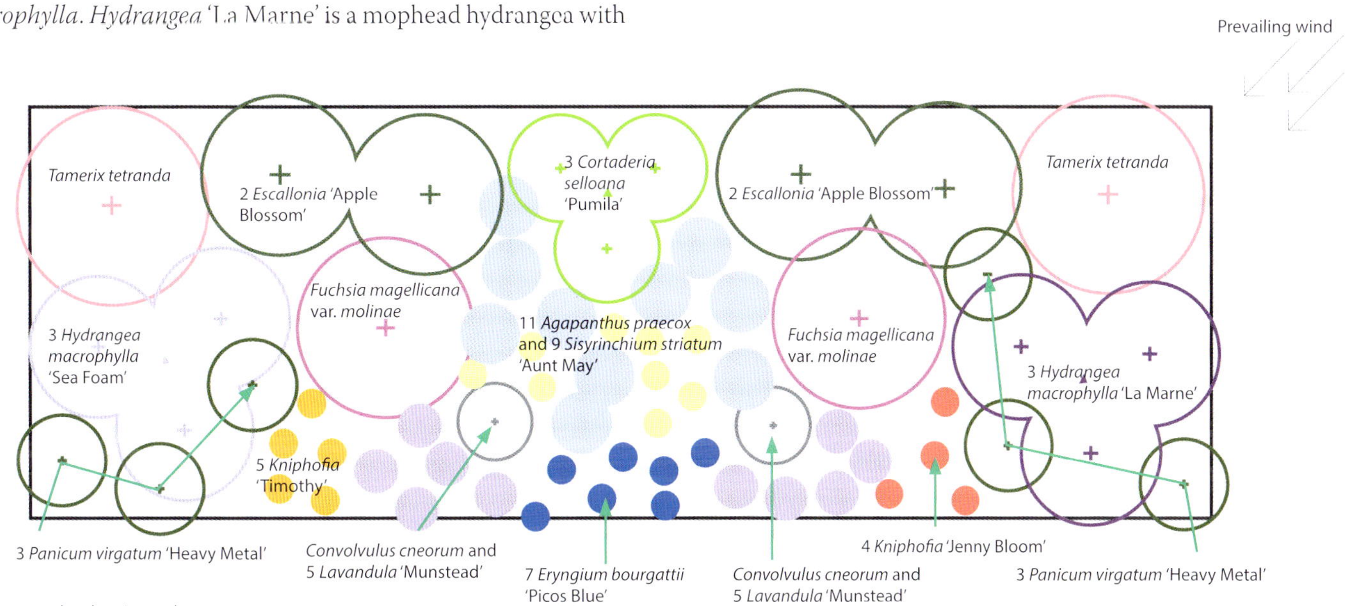

Pastels sample planting scheme.

Chapter 6

Biodiversity and Sustainability

DON'T FIGHT IT

A much-welcomed development in gardening over recent years has been the increased attention being paid to how wildlife friendly and how ecologically sustainable our gardens are. Gardens should be one of our important points of contact with the natural world and yet they have involved some rather environmentally damaging practices. A pristine lawn, for instance, has traditionally incurred a heavy carbon cost because of the fuel use of lawn mower and leaf blowers and the carbon cost of fertilisers. The use of pesticides has been also far too common, water use too high, and many landscaping materials are not sourced in a sustainable way. But things are starting to change. We have noticed a strong trend in our clients asking us to create wildlife-friendly areas of the garden, to use recycled landscaping materials, and to install rainwater harvesting systems.

The most important principle in the design of sustainable gardens is to work with the existing, natural conditions of your garden rather than fight them. In coastal areas soils are often sandy, poor, and free draining. It is most important that the plants that we use will tolerate these conditions to grow successfully. However, it is a good idea to feed well with a slow-release fertiliser such as blood, fish and bone annually, except for those plants such as lavender that prefer a poor soil and no fertiliser. It is important to water well during the first season and good practice to fill the planting hole with water and let it drain before the plant is put in place, but we do want to limit the amount of water we will use to sustain planting in the long term. Correct choice of plants is vital for this, and we will look at other helpful practices in water conservation later in this chapter.

To have the widest possible choice of plants in an exposed coastal garden, shelter from the wind is the main requirement and shelter belts and windbreaks are most important. However, until such planted screening is established, you can use netting or slatted fencing to filter the wind and avoid newly planted stock being scorched by wind or salt. A solid fence is usually not the best choice; instead of filtering the wind, a solid barrier creates eddies and swirls of wind which can be more damaging. A solid fence or wall will offer protection for a distance of about twice the height of the fence but then the damaging turbulent effects start.

A green-veined white butterfly (*Pieris napi*) feeding on purple loosestrife (*Lythrum salicaria*).

Wind tunnels can be created when the wind is funnelled along the side of buildings or reflected off walls and fences. At the end of such a wind tunnel there should be a shelter belt or windbreak of suitable shrubs that, once established, will be successful in filtering the wind without damage to planting on the other side of the belt. Windbreak netting or something similar can be used to protect a newly planted windbreak or shelter belt until it is established. More detail on creating planted shelter belts and windbreaks, as well as some windbreak planting plans, can be found in Chapters 4 and 5.

An important part of working with your garden environment rather than fighting it is also to accept, to a certain extent, the type of soil that the garden naturally has and to plant accordingly. But gardening inevitably involves at least some soil improvement and not only for reasons of fertility or workability. Soil types interact with the local climate to influence the microclimates of our gardens. Frosts are more likely over sandy soils than over soils with a higher clay content. However, sandy soils warm up more quickly in the spring and drain quickly. The addition of humus in the form of garden compost can help water retention. Clay soils become easier to work by the addition of compost and horticultural grit. Changing the nature of these soils requires large quantities of compost and grit to have a large effect. The alternative is to live with what you have.

As well as knowing your soil type it is important to know its pH, a measure of the soil's acidity. Soil test kits are available from garden centres and are easy to use. pH 7 is neutral, below 7 acidic and above 7 alkaline. If the garden is on limestone or chalk the soil will be alkaline. We do not recommend attempting to alter the pH of your garden soil; just be aware of what it is and of how it may affect your choice of plants. Hydrangeas can act as indicators as to what the pH is. Many coloured hydrangeas are blue on acidic soil and red or pink on alkaline ones.

BIODIVERSITY

There is a considerable amount of scientific research that points to gardens being important for wildlife and biodiversity. As well as supporting wildlife, a well-designed wildlife-friendly garden can be beautiful and fascinating. In the long term they can even be lower maintenance, as the diverse microhabitats, integrated with the local environment, become stable, balanced and less prone to weeds or pests. We discussed in Chapter 1 how you might design your garden to integrate with some common coastal environments.

Biodiversity is about creating a variety of habitats for a wide range of creatures. A garden that has good biodiversity will use a high percentage of native plants and may include some elements of habitat restoration. The planting of the right trees and shrubs can make it easier for birds to nest and insects to thrive. *Crataegus monogyna* is a good example. It can be grown as a tree or within a hedge and will cope with coastal conditions. The thorns act as a protection to nesting sites and the berries act as a food source. Over a hundred species of insects are dependent upon hawthorn for their life cycle. It is important that the garden attracts a range of insects as they are at the base of the food chain; a garden that has a wide range of insects will attract a good range of the higher forms of wildlife.

BIRD-FRIENDLY PLANTING

Birds are perhaps the most obvious and the most widely enjoyed garden wildlife. One drawback of a garden as a wildlife habitat is that they can be too isolated from their surrounding environment. Hopefully, as gardening for wildlife becomes more common and higher priority is given to the creation of 'green corridors' in urban planning, this will become less of a problem. Until then, the mobility of birds means that they can more easily access our

Siskin (*Carduelis spinus*), an increasingly common visitor to gardens particularly in the winter.

The mistle thrush (*Turdus viscivorus*) is larger and greyer than the more common song thrush.

gardens than many other sorts of wildlife. We can help encourage them to visit with our choice of planting and many of the plants for coastal gardens discussed in Chapter 5 are suitable for attracting birds into the garden.

Alnus cordata (alder) likes some moisture in the soil and the seed heads may well attract both siskin and redpoll in the winter. The siskin is an increasingly common visitor to gardens, particularly in the winter.

Cupressus macrocarpa may be used by goldcrests and firecrests for nesting purposes and *Sorbus aucuparia* is an easily grown native species. It flowers well and the berries are particularly attractive to blackbirds and mistle thrushes.

In a sheltered spot it is worth growing a blackberry. If weeds and grass are allowed to grow through the bush, warblers such as the chiffchaff may use it for nesting. Ivy is an essential in a wildlife-friendly garden if a sheltered, shady spot can be found. Wrens and robins may use it for nesting, and it flowers late in the year, providing an excellent source of nectar and later berries. It is a particularly good plant for invertebrates.

Climbers such as *Lonicera periclymenum* (honeysuckle) and *Rosa canina* (dog rose) will add interest to the native hedge as well as attracting a range of insects.

Hedgerows

Hedges consisting of several different cultivars can be especially wildlife friendly, providing nesting sites and food for birds, small mammals, and insects. At least 65 per cent of the hedge should be hawthorn and the rest can be made up of blackthorn (United Kingdom native), hornbeam (United Kingdom native), *Elaeagnus × ebbingei*, *Escallonia rubra* var. *macrantha*, *Griselinia littoralis*, *Olearia haastii*, *Berberis julianae*, and *Rosa rugosa*. If the garden is in a more sheltered spot away from the salt winds the remaining 35 per cent could consist entirely of United Kingdom natives such as field maple, beech, holly, and hazel. The most economical method of creating the native hedge is to purchase one-year-old slips and plant in a staggered double row at a rate of five to seven per metre, depending upon their size. After planting the height of the plant should be reduced to encourage them to grow bushier. If rabbits are likely to be a problem, then rabbit guards should be put in place. Once the hedge has become established *Rosa canina* (dog rose – United Kingdom native) and *Lonicera periclymenum* (honeysuckle – United Kingdom native) can be added to the planting. Such a hedge will provide nesting sites for a range of birds including dunnock, blackcap, blackbird, song thrush, and others.

Variation in height could be made a feature of the hedge by planting a few native trees within it. Suitable choices would be *Crataegus monogyna* 'Stricta' or if the hedge were a little less exposed to the salt winds *Carpinus betulus* 'Fastigiata' would be a good choice.

Rosa canina (this is the white form) is best grown in an established hedge. It will produce a good crop of hips in late summer.

Several of the shrubs included in the hedge can be an attractive feature on their own. Consideration can be given to *Berberis*, *Escallonia* and *Rosa rugosa*. Although not native, these three shrubs will add interest to the hedge. Other shrubs that are worth including in the coastal garden with birds in mind include *Ulex europaeus* 'Flore Pleno' (gorse – United Kingdom native) which provides nesting sites for whitethroats, *Ceanothus* 'Trewithen Blue' and *Fuchsia magellanica*. The ceanothus may attract pied wagtail, robin, wren and spotted flycatcher, especially if planted against a wall or fence, whilst the fuchsia may provide nest sites for linnets, yellowhammers and bullfinches.

Other food plants for birds

The use of a range of bird feeders is to be encouraged but the right choice of plants such as those discussed above will also provide a good food source. The seed heads on teasel will soon be found by goldfinches. The teasel is a biennial but self-seeds freely despite the finches. Rose hips are enjoyed by greenfinches and the berries on the guelder rose (*Viburnum opulus*) by bullfinches. This viburnum will require some shelter from the worst of the salt winds. Rough bark on mature trees is a source of insects for nuthatches, and conifers are attractive to the firecrest. This most attractive small bird is becoming more common in the United Kingdom.

BUTTERFLIES AND NATIVE PLANTING

Many birds feed on insects, and the planting and features within the garden should encourage as wide a range as possible. Butterflies add an extra dimension with their movement and colour. They will come to a wide range of flowering plants to feed on the nectar that they provide. Single flowers provide that nectar whereas double-flowered forms provide little or none. An early supply of nectar is required by the few species of British butterfly that overwinter as adults by spending the winter in a garden shed or in ivy. Only five species overwinter in this fashion: brimstone, red admiral, peacock, comma and small tortoiseshell. Early nectar can be provided by primrose, cowslip, wood anemone and aubretia. However, all of these will require shelter from the prevailing salt winds. For later in the year, *Buddleja davidii* and especially *Buddleja* × *weyeriana* 'Golden Glow', *Hebe albicans*, *Hebe* 'Midsummer Beauty', *Hylotelephium spectabile*, *Centranthus ruber*, *Lavandula* 'Munstead', *Thymus serpyllum*, *Erysimum* 'Bowles' Mauve', and Michaelmas daisy cultivars, e.g., *Aster* 'Barr's Pink' and *Aster frickartii* 'Monch', are excellent sources of nectar but would require some shelter from extreme exposure.

It is important to note that what this choice of plants may not achieve is the provision of food for caterpillars and that is vital if we are to achieve breeding colonies of butterflies within the garden. This food provision will come from native plants. If the garden contains a native hedge as previously suggested, then once it has matured to some extent, native plants can be planted at the base of it. These should be planted in groups of individual species rather than a spotty mix of totally different single species. However, some mixing should occur for the effect to look natural. Garlic mustard (*Alliaria petiolata*) – food plant for the green-veined white (*Pieris napi*) and the orange tip (*Anthocaris cardamines*) is a good choice. The green-veined white should not be confused with the large white (*Pieris brassicae*) and small white (*Pieris rapae*) which attack our brassicas. The primroses and wood anemone could be planted here along with some honesty (*Lunaria annua* – another food

A tree with rough bark is attractive to treecreepers (*Certhia familiaris*) which will start to explore the tree from the base, and the nuthatch (*Sitta europaea*), which will walk down the tree headfirst. Treecreepers usually nest behind pieces of loose bark whilst the nuthatch nests in holes in the trunk.

The nuthatch (*Sitta europaea*).

The firecrest (*Regulus ignicapilla*) is the United Kingdom's smallest bird and if it visits the garden it is most likely to be seen on conifers.

Self-sown seedlings of honesty (*Lunaria annua*) will flower the following year, will provide nectar, and are a food plant for the caterpillars of the orange tip and green-veined white butterflies.

Only the male orange tip has orange on its wings. The female has black.

plant for these two butterflies). Honesty can easily be grown from seed. Its seed pods are attractive and will be a source of seed for the following year. Hedge bedstraw should also be planted at the base of the hedge as it is the food plant for many moths.

Within the hedge there should be some native shrubs, as previously discussed. Blackthorn (*Prunus spinosa*) is the food plant of the black hairstreak (*Satyrium pruni*) which is not a common butterfly and would be a real bonus to have in the garden. Holly (*Ilex aquifolium*) could be in the hedge (requires some shelter) or elsewhere in the garden and is the food plant for the spring brood of the holly blue (*Celastrina argiolus*). A later brood feeds on ivy (*Hedera helix*).

Meadow geranium (*Geranium pratense*) and ox-eye daisy (*Leucanthemum vulgare*) are an attractive feature in the meadow. Native grasses play an important part as food plants for a number of butterflies.

WILDFLOWER MEADOWS

A wildflower meadow, with an emphasis on attracting butterflies and inevitably a host of other insects, is a fantastic feature, both for diversity and for its beauty. Ideally it should be in a sunny, sheltered spot. They are much harder to establish on a very exposed site. To create the meadow any existing turf should be stripped off along with any other vegetation, and all perennial weeds dug out. The soil should be of low fertility. If it is not, then some topsoil can be removed and used elsewhere in the garden. Rather than create the meadow from seed we prefer to use 'plugs', that is wild flowers that are sold in trays as small, already growing plants. They are planted at a rate of about seven per square metre. We can then choose exactly the species we wish to have in the meadow. It is very good practice to base your selection on what is found locally, and some suppliers will help you do this. That said, some of the chosen plants will perform very well, some will hold their own, and some will gradually die out. The appearance of the meadow will change from year to year. To attract as many butterflies as possible the meadow should be at its best in terms of flowering during the summer months. It should be cut towards the end of August to a height of about 7.5cm (3 inches); keep the sward low until the end of March. All cuttings should be left on the meadow for a couple of days to allow insects to return to the sward from the cutting. The cuttings should then be raked off and composted.

Our choice of plants should include native grasses, as several butterflies use them as food plants for their caterpillars. Creeping bent (*Agrostis stolonifera*) is the food plant for meadow brown (*Maniola jurtina*), small heath (*Coenonympha pamphilus*), and wall brown (*Lasiommata megera*). Cocksfoot (*Dactylis glomerata*) is the food plant for large skipper (*Ochlodes sylvanus*), Essex skipper (*Thymelicus lineola*), speckled wood (*Pararge aegeria*), wall brown, and meadow brown.

Yorkshire fog (*Holcus lanata*) is the food plant for small skipper (*Thymelicus sylvestris*), speckled wood and marbled white (*Melanargia galathea*). Yorkshire fog is an invasive species, so do not be tempted to plant it in flower borders. All these butterflies are likely to be attracted to the meadow in many areas of the United Kingdom.

Other plants that will provide summer colour in the meadow and should be included in our selection are lesser and greater knapweed (*Centaurea nigra and C. scabiosa*), field scabious (*Knautia arvensis*), lady's bedstraw (*Galium verum*), yarrow (*Achillea millefolium*), and kidney vetch (*Anthyllis vulneraria*).

The large, small and Essex skippers are likely to be seen in the meadow. The large skipper is identified by a thick black line through the centre of its upper wings.

The marbled white butterfly is unmistakeable and can be seen in July. It appears to have a liking for purple flowers.

Lady's bedstraw (*Galium verum*) is an excellent meadow plant and is also often found on sand dunes and shingle beaches. It is the food plant for the caterpillars of several moth species including four species of hawk-moth. Hawk-moth caterpillars can be identified by the hooked tail horn.

This caterpillar is of the lime hawk-moth (*Mimas tiliae*). This particular hawk-moth is not found on lady's bedstraw (its food plants are a number of trees including lime, alder and cherry cultivars) but the typical hooked tail horn of hawk-moth caterpillars is clearly visible.

Yarrow is the food plant for a range of moths, particularly pug moths. Examples include the common pug (*Eupithecia vulgate*), the lime-speck pug (*Eupithecia centaureata*) and the wormwood pug (*Eupithecia absinthiata*). The latter two are commonly found on salt marshes.

An area of the meadow could be cut more regularly to keep the sward to a lower level. Plants such as bird's foot trefoil (*Lotus corniculatus*), sheep's sorrel (*Rumex acetosella*) and red and white clover (*Trifolium pratense and T. repens*) should perform well in this situation. Bird's foot trefoil is an attractive wildflower and is a food plant for the clouded yellow (*Colias croceus*) and green hairstreak (*Callophrys rubi*). The presence of either of these butterflies in the garden would be another bonus.

The clouded yellow is mostly a migrant to southern coasts in the United Kingdom but visits most years. It has been seen over winter on cliffs in Hampshire and perhaps can now be considered as an all-year round resident. As a migrant, it normally comes from North Africa as do large numbers of red admiral (*Vanessa atalanta*) and painted lady (*Vanessa cardui*). Red admiral plus peacock (*Aglais io*), small tortoiseshell (*Aglais urticae*) and comma (*Polygonia c-album*) use stinging nettle (*Urtica dioica*) as their food plant. We do not want to introduce nettles into the meadow, but a patch of nettles elsewhere is a possible choice or grow them in a large container. The nettles must be grown in a sunny spot as it is fresh growth that the caterpillars require. Cut the nettles down at the end of winter and after the first batch of caterpillars have disappeared. It is possible to obtain two or three hatchings of caterpillars with proper maintenance of the nettles.

The Isle of Wight and the South Hampshire coast have small breeding colonies of the Glanville fritillary (*Melitaea cinxia*). The food plant is ribwort plantain (*Plantago lanceolata*). Although not the most attractive flower it is worth growing as it is a food plant for several species of moth as well. We doubt if you will be successful in having this fritillary visit your garden, but it is worth a chance if you are near existing colonies. On the edge of the meadow there could be a few native shrubs grown just to attract butterflies. Examples would be *Ulex europaeus* (gorse), *Cytisus scoparius* (broom) and *Rhamnus cathartica* (buckthorn) or *Frangula alnus* (alder buckthorn). The two buckthorns are the food plant of the brimstone (*Gonepteryx rhamni*); *R. cathartica* should be grown on alkaline soils and *F. alnus* on damp, acidic soils.

Yarrow is an excellent choice for the meadow. It attracts a range of insects but particularly moths.

Bird's foot trefoil is a plant for an area of the meadow that is kept cut to a lower level than the rest. It flowers from May to September. It is also the food plant for the common blue and silver-studded blue butterflies.

The clouded yellow is a migrant butterfly whose primary larval food plant is various species of clover.

The peacock butterfly is widespread in the United Kingdom. The eyespots on the wings are used to startle predators.

Small numbers of the red admiral overwinter as adults, but the population in the United Kingdom is boosted by large numbers of immigrants that originate from central Europe.

Formerly one of the United Kingdom's commonest butterflies the population of the small tortoiseshell appears to be in decline. A possible cause may be a parasitic fly *Sturmia bella*.

The Glanville fritillary is found mainly on the south coast of the Isle of Wight and an odd colony on the coast of south Hampshire.

Floral or pictorial meadows

An alternative to the wildflower meadow described above is a type known as floral or pictorial meadow. It will need to be in a sunny spot that is sheltered from the prevailing coastal winds. The meadow does not contain any native grasses, is seed raised and consists of annuals, perennials, or a mixture of both. The plants used will be both native and non-native. It is extremely important that the ground is well prepared before seed sowing. Perennial weeds should be removed, and annual weeds can be hoed on a sunny day and left to wither in the sun. No topsoil should be removed. The surface should then be raked to a fine tilth and left for a couple of weeks for annual weeds to germinate. These should be hoed as above. This process may require repeating.

The seed should be mixed well with dry, sharp sand and sown over the surface. The seed is sown at a rate of 3g per square metre (for annuals). The added sand is to ensure that the mixture is spread evenly. Do not rake in but firm the surface with a garden roller or even your feet to ensure good contact with the soil. If dry weather persists, apply a light watering to aid germination. Commercial seed mixes are available and will produce a colourful display through the summer and possibly early autumn. For annual seed mixes the process should be repeated the following year. For a perennial seed mix the process is similar and full details can be found on the internet.

The floral meadow is good for a range of pollinators but does not provide food plants for the caterpillars of many of the butterflies discussed under 'Wildflower Meadows'.

MOTHS AND A SHINGLE BEACH – PLANTING PLAN

Dungeness has a vast shingle beach famous for being the home of Derek Jarman. It is also famous for its moths, with over 600 species of macro-moth being recorded. We are not attempting to match that, but a shingle bed would be an interesting feature in our coastal garden. It is likely to attract several breeding moths with a planting of a good choice of larval food plants. It is worth pointing out that not all the moths mentioned in this section occur in all coastal regions of the United Kingdom. It is a matter of hoping that there are breeding colonies nearby that may spread into the garden. Luck may be on our side and one or two of the rarer species may arrive.

The planting plan requires a sunny spot and a well-drained soil. The site should have a large amount of shingle dug into the soil to mimic the shingle beach and after planting should have a layer of shingle, at least 6cm (2.5 inches) thick, spread on top. Good drainage is an absolute essential for this design. The moths mentioned here are mainly found on plants growing in the shingle and/or neighbouring sandhills or cliffs.

The *Tamarisk tetrandra* in the top right-hand corner of the design, along with the broom (*Cytisus scoparius*) and gorse (*Ulex europaeus* 'Flore Pleno'), will act as a windbreak although most of the plants in the design do not require one. Tamarisk is the larval food plant for the Channel Islands pug (*Eupithecia ultimaria*) which is a recent colonist and immigrant along the coast of Southern England, the Isle of Wight, and the Channel Islands.

Even a small area of floral meadow makes an attractive, colourful feature. Cornflowers (*Centaurea cyanus*) feature prominently in this one.

Local authorities are increasingly planting floral meadows to enhance their area. They help to increase the biodiversity along the roadside verges.

Broom is the larval food plant for light brocade (*Lacanobia w-latinum*), the streak (*Chesias legatella*) and oak eggar (*Lasiocampa quercus*) moths. The male oak eggar flies during the day. July belle (*Scotopteryx luridata*) occurs on shingle beaches, as well as other areas throughout much of the United Kingdom, and uses gorse as its food plant.

Sea kale (*Crambe maritima*) is a major feature of the planting in the design. It is a spectacular plant which starts to appear in growth in late March and by the end of May masses of small, fragrant white flowers appear. These attract a wide range of pollinators such as bees, hoverflies, and moths.

The planting of wild carrot (*Daucus carota*) and valerian (*Centranthus ruber*) will add summer colour to this bed. The wild carrot is a biennial and a good plant for wildflower meadows. There is a purple or reddish flower in the centre of the umbel which performs the function of attracting insects. Red valerian is often visited by the hummingbird hawk-moth (*Macroglossum stellatarum*) which is an immigrant but can occur in large numbers during periods of warm weather. In some years it may breed in the United Kingdom. The larvae feed on lady's bedstraw (*Galium verum*). Wild carrot is the food plant for yellow belle (*Aspitates ochrearia*) which occurs in all coastal counties in Southern England and Wales. Another food plant for the yellow belle is buck's-horn plantain (*Plantago coronopus*). The related sea plantain (*Plantago maritima*) is the food plant for the extremely attractive belted beauty (*Lycia zonaria*) which is often found near bird's foot trefoil (*Lotus corniculatus*) on warm afternoons. The Jersey tiger (*Euplagia quadripunctaria*) also flies on warm days and often visits *Buddleja davidii*. It has several larval food plants including plantains. Until recently the Jersey tiger was restricted to the Channel Islands and parts of the south coast of England. It is now found in an increasing number of counties of the United Kingdom.

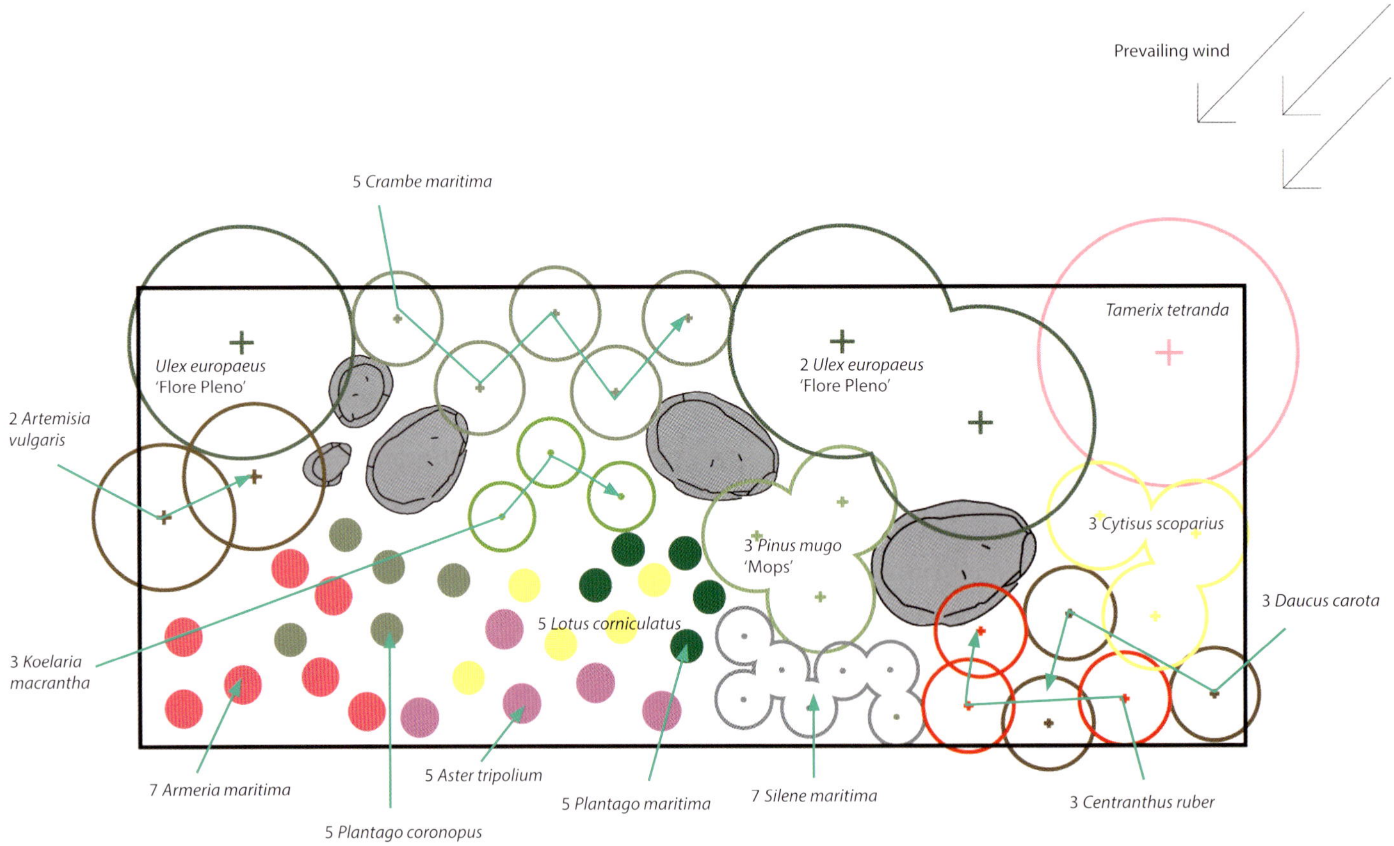

Planting plan of shingle bed for moths.

Sea kale (*Crambe maritima*) in its natural habitat; it is also a good garden plant for well-drained areas of the coastal garden.

The hummingbird hawk-moth is an attractive day-flying immigrant moth.

There are two generations of the yellow belle in each year. They occur in early and late summer.

The Jersey tiger moth is very localised in the United Kingdom and is found most frequently along the south coast.

Bird's foot trefoil, which does require some protection from the coastal conditions, may also be visited during the day by the six-belted clearwing (*Bembecia ichneumoniformis*) and six-spot burnet moths (*Zygaena filipendulae*) for which it is the food plant.

Behind the plantains is some crested hair grass (*Koeleria macrantha*) which prefers alkaline soils and could be visited by several moths that use grasses as food plants, examples being the dusky sallow (*Eremobia ochroleuca*) and the hedge rustic (*Tholera cespitis*). The hedge rustic, despite its name, is not a moth of the hedgerow but is common on grassland and in gardens. Occasionally it does occur on the coast.

Pinus mugo 'Mops' is a slow growing, dwarf cultivar and adds a different dimension to the planting. It may be wishful thinking that it will attract the pine hawk-moth (*Sphinx pinastri*), the larvae of which can be found in summer on both young and mature pines but prefer old needles. It is usually found on the Scots pine (*Pinus sylvestris*). Will it use a dwarf *Pinus mugo* as the food plant? Possibly, if stands of Scots pine are close by. Along the front of the planting, thrift (*Armeria maritima*) and sea campion (*Silene maritima*) will provide colour in spring and early summer and the sea aster ((*Aster tripolium*) will flower later. Between them we would hope to attract several moths including the scarce thrift clearwing (*Synansphecia muscaeformis*), star-wort (*Cucullia asteris*) and feathered brindle (*Aporophyla australis*). The star-wort mainly occurs along the eastern and southern coasts of England and its larvae feed on the sea aster (*Aster tripolium*). The feathered brindle has the sea campion (*Silene maritima*) as its food plant and is found along the southern coasts of England and South Wales.

Other moths may be attracted to our version of a shingle beach by planting the right food plants. Starry pearl (*Cynaeda dentalis*) is restricted to a few localities in southeast England. Its food plant is viper's bugloss (*Echium vulgare*).

The toadflax brocade (*Calophasia lunula*) occurs on shingle beaches along the south-eastern and southern coasts of England. But its range does appear to be expanding. Its food plants are purple and common toadflax (*Linaria purpurea* and *Linaria vulgaris*). The larvae are very colourful but can be difficult to spot. Plant the food plants with care as they are invasive.

The larvae of the six-belted clearwing feeds on the roots of its food plant.

The six-spot burnet flies with a slow buzzing flight on sunny days.

The thrift clearwing may be seen on the flowers of thrift and thyme. The larvae feed inside the roots and stems of these food plants.

The feathered brindle is not common but may be found along the coast of England from Norfolk to Cornwall and is particularly found on the shingle at Dungeness.

Starry pearl (*Cynaeda dentalis*) is a very distinctive moth but is only found in a few coastal localities.

The toadflax brocade has very colourful larvae.

Viper's bugloss (*Echium vulgare*) is commonly found on sand dunes.

Lichens on a range of surfaces are a sign of clean air. The caterpillars of several moths use lichen as a food supply.

The pigmy footman is a scarce species, only occurring in a few coastal localities.

The dew footman is easily recognised by its colour (cream to yellow) and its markings.

But why the rocks?

We have included some rocks in the planting plan for a moth-attractive shingle beach style planting. Lichens can be found on most surfaces within the garden such as trees, rocks, and paving. Cleaner air and reduced quantities of acid rain have led to an increase in the number of lichens growing and a corresponding increase in the numbers of some of the moths whose caterpillars feed on them. The main group of species are the footman moths. It would be interesting to have lichens on the rocks in our coastal garden and to see which species, if any, arrive. Pigmy footman (*Eilema pygmaeola*), scarce footman (*Eilema complana*) and hoary footman (*Eilema caniola*) are possibilities although they are not common moths. The pigmy footman is found on sand dunes and shingle beaches and occurs in both Kent and Norfolk. The scarce footman is found in coastal locations in both Wales and Ireland but tends to be on heathland in England. The hoary footman is found on seaside cliffs and shingle beaches mainly on the southwest coasts of England and Wales.

Other species that may arrive include the dew footman (*Setina irrorella*) and the marbled green (*Nyctobrya muralis*). Males of the dew moth fly in the afternoon, as well as at night. It is scarce but has a scattered distribution along the southern coasts of the United Kingdom. The marbled green is a beautiful moth which is also confined to southern coasts. All these moths are on the wing during summer months, especially July and August.

The growth of lichen on the rocks can be encouraged by painting the surface of the rock with a weak yoghurt solution especially on the shadier side of the rocks. If no signs of lichen growth are seen after two weeks, repeat the process. It may take a couple of years to achieve really good results.

ATTRACTING MOTHS AND OTHER INVERTEBRATES

Butterflies and moths are desirable invertebrates that make the garden more interesting than if they were absent. There are, however, a huge number of other invertebrates that will increase the biodiversity of the garden. They will range from those that are the gardener's friend, those that are neutral in terms of the health of plants, and a few that may cause the gardener some problems. That said, invertebrates are essential if the garden is going to be healthy. They are the start of the food chain, they are important pollinators, and they are a natural control on those that the gardener may consider to be pests. The days of spraying pesticides, herbicides and fungicides should stop. The garden is not a place for poisons and their use will reduce the biodiversity within it. We have discussed the features of a native hedge, a wildflower meadow, a shingle beach, and a patch of nettles in terms of their attraction for both moths and butterflies. We shall now look at how these features can attract a greater range of invertebrates than just the butterflies.

Nettles, as well as being a food plant for the larvae of the red admiral and other butterflies, are also a food plant for several colourful moths such as the ruby tiger (*Phragmatobia fuliginosa*). This moth is common in various habitats including gardens and sand dunes. The larvae feed on several different herbaceous plants including nettles and plantains. The scarlet tiger (*Callimorpha dominula*) can also be found in coastal habitats, including the upper parts of the beach, mainly in southwest England and Wales. It is a very colourful moth and the male flies during the day as well as at night. The larvae feed on nettles, honeysuckle, and other plants. The garden tiger (*Arctia caja*) appears to be in decline but does occur on sand dunes. Its larvae feed on a variety of plants, including nettles, and are very hairy. They are known as 'woolly bears' and handling them can irritate the skin. Beware.

Another colourful moth is burnished brass (*Diachrysia chrysitis*) which is common in a wide variety of habitats including the garden. Its food plants include nettles and thistles. The dot moth (*Melanchra persicariae*) is widespread in the United Kingdom. Its larvae feed on a variety of wild and garden plants including nettles. It is important to note that the patch of nettles should be sited in a sheltered, sunny spot. Because the nettle is an invasive plant, consider planting it in a large container or small raised bed.

If deciduous trees such as alder (*Alnus cordata*), white willow (*Salix alba*), hornbeam (*Carpinus betulus*) and rowan (*Sorbus aucuparia*) have been planted as part of a windbreak or, even better, in a sunny, more sheltered spot, then they may act as the

The scarlet tiger can be seen on the wing in the day.

The larvae of the scarlet tiger are very colourful and occur on a range of plants.

The caterpillar of the dot moth (*Melanchra persicariae*) is highly likely to be seen in the garden on a variety of plants.

When at rest, the buff-tip (*Phalera bucephala*) resembles the broken twig of a birch tree. It is a superb example of camouflage.

food plant for the beautiful buff-tip (*Phalera bucephala*). The caterpillars are in quite large groups and can defoliate young trees, but the damage is not usually terminal. They can be moved to a larger tree if the gardener is concerned. The number of caterpillars will be reduced by predators. The caterpillars have been recorded on both hawthorn and blackthorn, so it is worth looking at the hedges.

J.B.S. Haldane, a British evolutionary biologist, is often quoted as saying that our creator 'must have had an inordinate fondness for Beetles'. His quote is based on the fact that there are more than 400,000 species of beetle in the world, which represents about 40 per cent of all insect species. Although some species of beetle in the United Kingdom receive a 'bad press' e.g., lily beetle (*Lilioceris lilii*) and vine weevil (*Otiorhynchus sulcatus*), many are very useful within the garden. *Demetrias atricapillus* is a reddish-orange ground beetle which is fairly common in the southern half of England. It is found in grassland but appears to be more coastal (sand dunes) in the west and in Wales. It feeds mainly on aphids. *Ceutorhynchus pollinarius* is a weevil whose larvae feed on the roots of nettles. The nut weevil (*Phyllobius pomaceus*) is common throughout England and Wales and occurs in various habitats including gardens. The adult is bluish-green when fresh and usually found on nettles; the larvae feed on the roots.

Ladybirds are beetles and are great friends of the gardener. Many of the ladybirds found in the United Kingdom are feeders on aphids both as adults and larvae. An adult ladybird will consume *c.*5,000 aphids in its lifetime. The commonest ladybirds in the garden are the 2-spot and the 7-spot. The 11-spot ladybird (*Coccinella undecipunctata*) is found in various habitats but especially those by the sea. The number of spots varies from seven to eleven and it is commonly found on nettles, thistles, and gorse. It also feeds on aphids.

Weevils form one of the largest families of beetles and can usually be recognised by their antennae which have an obvious elbow. The vine weevil (*Otiorhynchus sulcatus*) is a major pest in British gardens. The larvae are mainly active from summer to spring, feeding on the roots of a range of plants but particularly primulas, heucheras and fuchsias. They will quite commonly attack container-grown plants.

Unlike the vine weevil, the nut weevil is not a problem in the garden.

The 11-spot ladybird, like most other ladybirds, is a voracious predator of aphids.

The common green capsid can be found throughout the summer.

Various bugs are likely to be found on nettle patches. The common green capsid (*Lygocoris pabulinus*) is a common sap-sucking bug found on various plants but mainly nettles. It can be found from June to October. The adult of the common nettle capsid (*Liocoris tripustulatus*) can be found all year. Nettle is the food plant for all stages of the insect.

As well as a patch of nettles it is worth giving some thought to having a patch of thistles. They too can spread and it may be worth having them planted in a container or raised bed. They do self-seed and if this is likely to be a problem the flower heads can be removed as the flowers fade. The hoary bell (*Eucosma cana*) is a common moth which uses various thistles, *Cirsium* and *Carduus* species, as the food plant. The larvae feed on the flower heads. The ermine knot-horn (*Phycitodes binaevella*) is common throughout the United Kingdom and is easily identified whereas other species of *Phycitodes* are not. *P. binaevella* has prominent black dots on a pale ground. The larvae feed on the flower heads of spear thistle (*Cirsium vulgare*). The common yellow conch (*Agapeta hamana*) is another easily identified moth. It has brown markings on a pale yellow ground colour and is common throughout the United Kingdom. The larvae feed on the roots of various species of thistle. The thistle bell (*Epiblema scutulana*) is a moth that is difficult to identify to species level. It is a common moth whose larvae feed on the roots of spear thistle and musk thistle (*Carduus nutans*). It must not be forgotten that as well as the moths the migrant butterfly painted lady (*Vanessa cardui*) also uses thistles as its food plant.

Lema cyanella is a small, metallic greenish-blue beetle that occurs in various habitats including salt marshes throughout England and Wales. The larvae feed on the leaves of various thistles, often covering themselves in excrement as a method of protection from predators. *Crepidodera impressa* is a rare leaf beetle that has been found on the coasts of Hampshire, Dorset, and Devon. Marsh thistle (*Cirsium palustre*) is a possible food plant. Three species of tortoise beetle are found associated with thistles. The thistle tortoise beetle (*Cassida rubiginosa*) is common in both England and Wales and both adult and larvae feed on the leaves of various plants including thistles. *Cassida vibex* is quite common in Southern England where west country records are mainly coastal. It has several food plants but especially thistles. The larva feeds on leaves and covers itself in excrement. It is often found with *Cassida rubiginosa*. The green tortoise beetle (*Cassida viridis*) is widespread in England and Wales but it can be found in coastal areas on dead nettles (*Lamium species*) at the bottom of hedges. It is easily identified as it is all green. Its larvae do not cover themselves in excrement.

The planting in the shingle bed was chosen with moths in mind but if the bed is of a decent size, it could attract a range of interesting invertebrates that are attracted to coastal areas. This is more likely to occur if there are existing colonies in the area in question. The boat bug (*Enoplops scapha*) is a shieldbug that occurs on dry cliffs and dunes along the south coast of England and Wales. It feeds on the leaves of a range of different plants, including wild carrot and wild thyme. The gorse shieldbug (*Piezodorus lituratus*) is large and common throughout the United Kingdom. It is not confined to coastal areas but occurs where gorse and broom are growing.

The thistle tortoise beetle is the most common tortoise beetle in the United Kingdom.

The green tortoise beetle is entirely green and lacks the markings of similar species.

Schilling's rhopalid (*Chorosoma schillingi*) is a bug with long legs and antennae. It is mainly southern and coastal, and is found from Norfolk to Anglesey. It feeds on various grasses and is beautifully camouflaged as it walks though the grass.

Inevitably the gorse (*Ulex europaeus*) will be the food plant for several weevils. The gorse seed weevil (*Apion ulicis*) is common in England and Wales. It is not confined to coastal areas but is likely to be found there where gorse is thriving. It is a small weevil and the larvae feed on the seeds of its host plant. *Polydrusus confluens* is a widespread weevil along the coasts of Southern England and Wales. Its food plant is gorse, the adult feeding on new shoots and the larvae on the roots. It does seem to prefer the dwarf form *Ulex minor*. Other plants in the shingle bed may also attract weevils. *Sitona puberulus* is a small weevil usually found in coastal areas on bird's foot trefoil (*Lotus corniculatus*). It reaches as far north as southwest Scotland but is rather localised. *Ethelcus verrucatus* is a rare weevil which occurs along the south coast of England from Suffolk to Cornwall. It is only found on the roots of the yellow horned poppy (*Glaucium flavum*) which would need to be added to the above planting plan. It would be exciting if it were to colonise the garden. A similar comment could be made if a couple of rare weevils were to appear on the planting of sea kale (*Crambe maritima*). *Ceutorhynchus cakilis* is a rare weevil which feeds on the leaves and the larvae live in the fruits, feeding on the seeds, before pupating in sand. The larvae of the cabbage leaf weevil (*Ceutorhychus contractus*) feed on brassicas as well as other plants. It occurs in the southern half of England and in Wales, and is mainly a coastal species.

The sulphur beetle (*Cteniopus sulphureus*) is not a weevil but a small beetle. It is easily recognised as it is the only bright yellow beetle. It has a southerly distribution in the United Kingdom and is almost entirely coastal. It is often found on sand dunes on wild carrot and wild thyme.

Tiger beetles (*Cicindelinae* spp.) are a large group of beetles of over 2,000 species worldwide. Only five species occur in the United Kingdom, of which four are rare. The green tiger beetle (*Cicindela campestris*) is quite common and can also be found on sand dunes and nearby sandy grassland. It is recognised by its metallic green colour with pale spots. It nests in holes in sandy soil where the larvae stay over winter. If the garden is close to the sea and has some sandy patches in the lawn the adult may be seen between April and September. It is a ferocious predator of spiders, ants and caterpillars and can run at speeds of up to 5mph (8kmph), making it one of our fastest moving insects.

Hoverflies are extremely important pollinators and as gardeners we should do as much as possible to encourage them into the garden. Many of their larvae feed on aphids. Many species look like small wasps, but do not sting, and will be seen hovering above flowers. There are several species that will be happy in coastal conditions. The marmalade hoverfly (*Episyrphus balteatus*) is a possible migrant as it is often found in large numbers in coastal areas when it will feed on coastal flowers. It is our commonest hoverfly and is found in most areas of the United Kingdom. The larvae feed on aphids and the adults on nectar from garden flowers.

The boat bug is a distinctive shield bug usually found on coastal cliffs.

The wings of *Chorosoma schillingi* reach halfway down the abdomen.

The sulphur beetle is the only bright yellow beetle found in the United Kingdom.

The marmalade hoverfly can be seen throughout the year.

The large spotted-eyed dronefly is on the wing in spring and early summer.

The large spotted-eyed dronefly (*Eristalinus aenus*) is a spring species and can be seen on the wing from April to June. It is often seen on the flowers of blackthorn (*Prunus spinosa*) or alexanders (*Smyrnium olusatrum*). Although it can be found inland it is mainly a species of the saltmarsh and coastal areas. The larvae often feed on rotting seaweed. If the eyes of the adult are separated on the top of the head, it is a female; if they meet in the middle, it is male. In both, the eyes are very spotty!

BEES

If there is a wildflower meadow in the garden, then as discussed earlier, there will be a greater number of butterflies in the garden. But that is only part of the story because it will be an ideal habitat for a large range of invertebrate species, particularly if the meadow is in a sunny spot that is sheltered from the salt winds. One group of invertebrates not yet discussed will benefit greatly from it, and that is bees. Bees contribute a huge amount to man's welfare as they are major pollinators and play an especially important part in the production of our food. Gardeners should do as much as is possible to encourage them within the garden. Honeybees are attracted to a whole host of flowers, particularly those that are coloured blue or violet. However, they do sting and having hives within the garden may be a problem if young children are present. But all is not lost, for there are many species of bees and the majority do not sting.

The United Kingdom has twenty-three species of bumblebee. They do not make honey. Nests of bumblebees can be found in dry tussocks of dead grass or are underground, often in disused mouse burrows. Their larvae survive by feeding on stored pollen, especially pollen from clovers, peas, beans and vetches, i.e. legumes. Six of the species – the 'big six' – are common and found over most of the United Kingdom. They are likely to be in our coastal garden. They are the buff-tailed bumblebee (*Bombus terrestris*), the white-tailed bumblebee (*Bombus lucorum*), the red-tailed bumblebee (*Bombus lapidarius*), the garden bumblebee (*Bombus hortorum*), the early bumblebee (*Bombus pratorum*) and the common carder bumblebee (*Bombus pascuorum*). The plantings suggested so far, plus that of the wildflower meadow, are likely to supply the requirements of the 'big six'. Bumblebees sometimes concentrate on a group of a single species of flower when foraging. This is known as constancy. Consequently, when planting for bumblebees it is better to plant large groups of the same flower together, rather than one of everything. Good choices of plant include rosemary (*Rosmarinus officinalis*) and cultivars of old English lavender (*Lavandula angustifolia*) such as 'Twickel Purple'.

Agastache foeniculum and *Monarda* 'Mohawk' are also good choices and especially important are bird's foot trefoil, red clover and wild broom. Many of these would need to be in a sheltered, and sunny spot away from the prevailing wind. Plants that will withstand coastal winds and salt to a reasonable extent (*see* plant list in Chapter 5) and are good for bumblebees include broom, rosemary, *Buddleja globosa*, *Buddleja* × *weyeriana* 'Sungold', *Ceanothus*, *Escallonia*, *Rosa rugosa*, thyme, borage (annual), cornflowers (annual), marjoram, wild scabious, *Eryngium* and thistles. Rare bumblebee species are more specialised and require large areas of forage, particularly bird's foot trefoil, red clover and devil's-bit scabious.

The common carder bee is on the wing from March to November.

The shrill carder bee is one of our rarest bumblebees and has a loud, high-pitched buzz which gives it its name.

Carder bumblebees nest in long vegetation such as tussocky grass. Four of the species are rare and only one is widespread in the United Kingdom, namely the common carder bee (*Bombus pascuorum*). The brown-banded carder bee (*Bombus humilis*) is rare and is mainly found in coastal areas in Southern England and Wales but is also found in a few inland sites. It is on the wing from mid-May onwards. It is all ginger with a darker brown band on the abdomen. The moss carder bee (*Bombus muscorum*) is also all ginger but has no black hairs at all on its body. It is widespread in the United Kingdom but outside of Scotland and Ireland is mainly coastal, particularly in the south of England. The shrill carder bee (*Bombus sylvarum*) is probably our rarest bumblebee and is an endangered species. It is straw coloured with a broad black band on the thorax and a reddish-orange tail. It occurs in a few sites in South Wales and Southern England and is found inland as well as on the coast.

Cuckoo bumblebees are parasites. A female will enter the nest of a social bee i.e. one that forms a colony with a single queen. It kills the queen and establishes itself as the queen and is waited upon by the host workers. She will lay her eggs in the nest and leave them to be looked after. As the cuckoo bee does not feed the larvae, they have no pollen baskets on their legs. The southern cuckoo bee (*Bombus vestalis*) is a widespread species particularly in Southern England. It is not restricted to the coast. The red-tailed cuckoo bee (*Bombus rupestris*) is a large bee and is a parasite of the red-tailed bumblebee (*Bombus lapidarius*). It is widespread across southern England.

Unlike social bees, solitary bees have no communal nest. Most of the year species survive as larvae that grow in their cells eating the store of pollen that has been left for them. There are several that are likely to be found in the garden. The hairy-footed flower bee (*Anthophora plumipes*) is widespread across the United Kingdom and nests in tunnels in dry ground or in old mortar in walls. It flies from late February to mid-June and has a liking for early flowering comfrey such as *Symphytum grandiflorum* 'Hidcote Blue'. The *Symphytum* will object to the salt winds and is happier when planted in a sheltered spot in sun or shade. Also found nesting in old mortar as well as sandy embankments is the red mason bee (*Osmia rufa*). It is common in gardens throughout England, Wales, and southern Scotland. It is an excellent pollinator. Most gardeners come across semi-circular pieces missing from the leaves of their roses. The bee responsible is the leaf cutter bee (*Megachile* spp.), of which there are seven species in the United Kingdom. The pieces of leaf are used to wall up the cells which are in tunnels or tubes provided by the gardener. They are also excellent pollinators. They often visit knapweeds (*Centaurea* spp.) which can be planted in the wildflower meadow. If the lawn is cut short and has some bare patches in it, then the gardener may observe some volcano-like holes of earth. This is the work of the tawny mining bee (*Andrena fulva*) which makes a tunnel in the short grass in which to lay the eggs. It is widespread across England and Wales and is an excellent pollinator, particularly of fruit trees, but of many other plants as well. They are totally harmless and will not sting.

The solitary bees discussed above occur widely across inland Britain but there are species that can be considered as concentrating on coastal habitats. One such example is the black mining bee (*Andrena pilipes*). This occurs from the Thames Estuary to Cornwall where it nests in burrows in light sandy soil on landslips, clifftops and sand dunes. Pollen is collected from brassicas (such as sea kale) and thistles, amongst other plants. Consequently, it may be seen in coastal gardens if they are near to an existing colony and these plants are in the garden.

The tawny mining bee is one of the United Kingdom's most common solitary bees. It is very distinctive and easily recognised.

The black mining bee (*Andrena pilipes*) nests in sandy soil on land slips and cliff tops.

The long-horned bee (*Eucera longicornis*) is an uncommon bee found on coastal grassland and soft cliffs. It is confined to south Devon, Cornwall, and south Wales and appears to be attracted to vetches and other legumes. The female pantaloon bee (*Dasypoda hirtipes*) is easily recognised by the oversized pollen brushes on its hind legs. It is a mining bee and nests in sandy soils, often in large groups, usually in coastal and heathland habitats. It is common on both the southeast and southwest coasts of England as well as west Wales. It appears to prefer to forage on yellow flowers such as ragwort (*Senecio vulgaris*), fleabane (*Pulicaria dysenterica*), and common cat's ear (*Hypochaeris radicata*). The leaf cutter bee (*Megachile maritima*) is found on coastal sand dunes on the English coast but seldom in the northeast. It is attracted to sea holly (*Eryngium maritimum*) and bird's foot trefoil. The latter does provide a rich source of food for a variety of bees.

Designers Wendy Booth and Leslie Howell used *Santolina chamaecyparissus* (cotton lavender) and *Centranthus ruber* (red valerian) to form an effective backbone to the planting around this coastal pond.

PONDS

A garden pond is a haven for wildlife and is highly recommended as a feature that will increase the biodiversity of the garden. It will be used for drinking and will be a habitat for breeding by a range of aquatic wildlife such as diving beetles, pond skaters and a host of smaller organisms. Frogs, toads, and newts are all likely to use it.

Ponds can be purchased preformed, or (for larger ponds or those with more complicated shapes) constructed from butyl liners. At least part of the pond should be in full sun and ideally at least 2.5 feet deep (0.75 metres) in order to avoid it freezing solid in the winter. Fish should not be added to the wildlife pond as they are top predators and contribute to nutrient levels within the pond. This encourages the growth of algae and blanketweed.

The pond should be at two levels, one consisting of shelves for container planting of marginal plants, and deeper water for plants such as water lilies. The marginal plants can be irises such as *Iris ensata*, flowering rush (*Butomus umbellatus*) and marsh marigold (*Calustris palustris*). Bulrush (*Typha latifolia*) is not recommended for small ponds as it is invasive and may well puncture the pond liner. The yellow flag iris (*Iris pseudacorus*) is also a vigorous grower and needs planting with care. Oxygenating plants should also be added. Recommended oxygenators are hornwort (*Ceratophyllum demersum*) and water starwort (*Callitriche autumnalis*). It is important that wildlife can get in and out of the pond easily – and this includes hedgehogs, which do have a habit of falling in.

The southern hawker is a very inquisitive species, often approaching a couple of feet from the observer.

A seat by the pond is an essential as watching dragonflies and damselflies is one of the pleasures that a wildlife pond can bring to the garden. They do require fresh water for breeding although a few species will breed in brackish water e.g. the southern emerald damselfly (*Lestes barbarus*) which is a recent coloniser of the United Kingdom. Damselflies fold their wings back along the body when at rest whereas dragonflies hold their wings at right angles to the body. Fresh water pools and streams that are close to the coast, or those of low salinity, often attract good numbers of dragonflies and damselflies and they are likely to be seen in the garden. They will fly quite large distances in order to feed.

Dragonflies that may well visit the garden include the southern hawker (*Aeshna cyanea*), common darter (*Sympetrum striolatum*), emperor (*Anax imperator*), beautiful demoiselle (*Calopteryx virgo*) and common blue damselfly (*Enallagma cyathigerum*).

The common darter has the habit of returning to the same perch which it may occupy over several days.

The emperor dragonfly favours larger ponds, keeping well away from the banks.

This damselfly, beautiful demoiselle, favours slow-moving streams. It has a slow, fluttering flight.

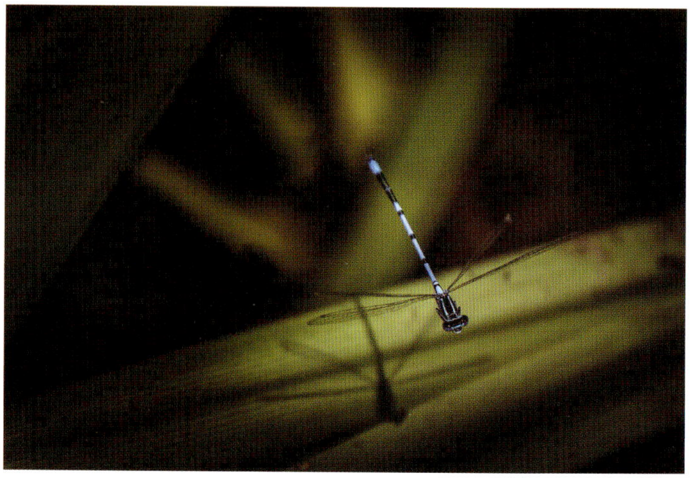

The common blue damselfly favours ponds with plenty of marginal vegetation. Usually flies low to the water.

PLANTING PLAN TO ENCOURAGE INSECTS

If your garden is not large enough to contain all or any of the features covered in this chapter then it is worth considering this planting scheme, the aim of which is to attract a broad cross-section of the insects that have been discussed. It needs to be in well-drained soil in a sunny spot with some shelter from the prevailing coastal wind.

Close to this planting would be an ideal site for a 'bug hotel'. Bug hotels can be a good place for solitary bees to live, breed and hibernate. They are also good for lacewings and ladybirds which are major devourers of aphids. Other insects will also find the hotel to their liking. Bug hotels can be purchased or made from recycled pallets. Instructions for making them can be found online. The spaces in the hotel should be filled with a range of different materials such as moss, dry leaves, pieces of brick, bark, pinecones, hollow canes, and similar materials, preferably of an organic nature. Topping the structure with old roof tiles will help to keep the hotel dry.

A log pile will also be useful. This should be in a damp shady area. Gaps between the logs can be filled in a similar manner to the bug hotel. The pile will be good for centipedes, woodlice, and beetles, particularly those that breed in rotting wood such as stag beetles (*Lucanus cervus*) and the wasp beetle (*Clytus arietis*). Stag beetle larvae may spend up to six years in the decaying wood. The wasp beetle lays its eggs in crevices in the bark of a wide range of fallen trees. The larvae then develop within the dead wood. They have a two-year life cycle. Both these beetles are easily identified.

The bug hotel should be sited in a sheltered part of the garden away from the prevailing wind. Solitary bees prefer a sunny spot whilst beetles and amphibians prefer it to be slightly damp.

Achillea 'Terracotta' has apricot-coloured flowers which fade to creamy yellow. It is very attractive to pollinators, especially hoverflies.

The wasp beetle (*Clytus arietis*) is active from spring to early summer. It is one of the most common European longhorn beetles and is totally harmless.

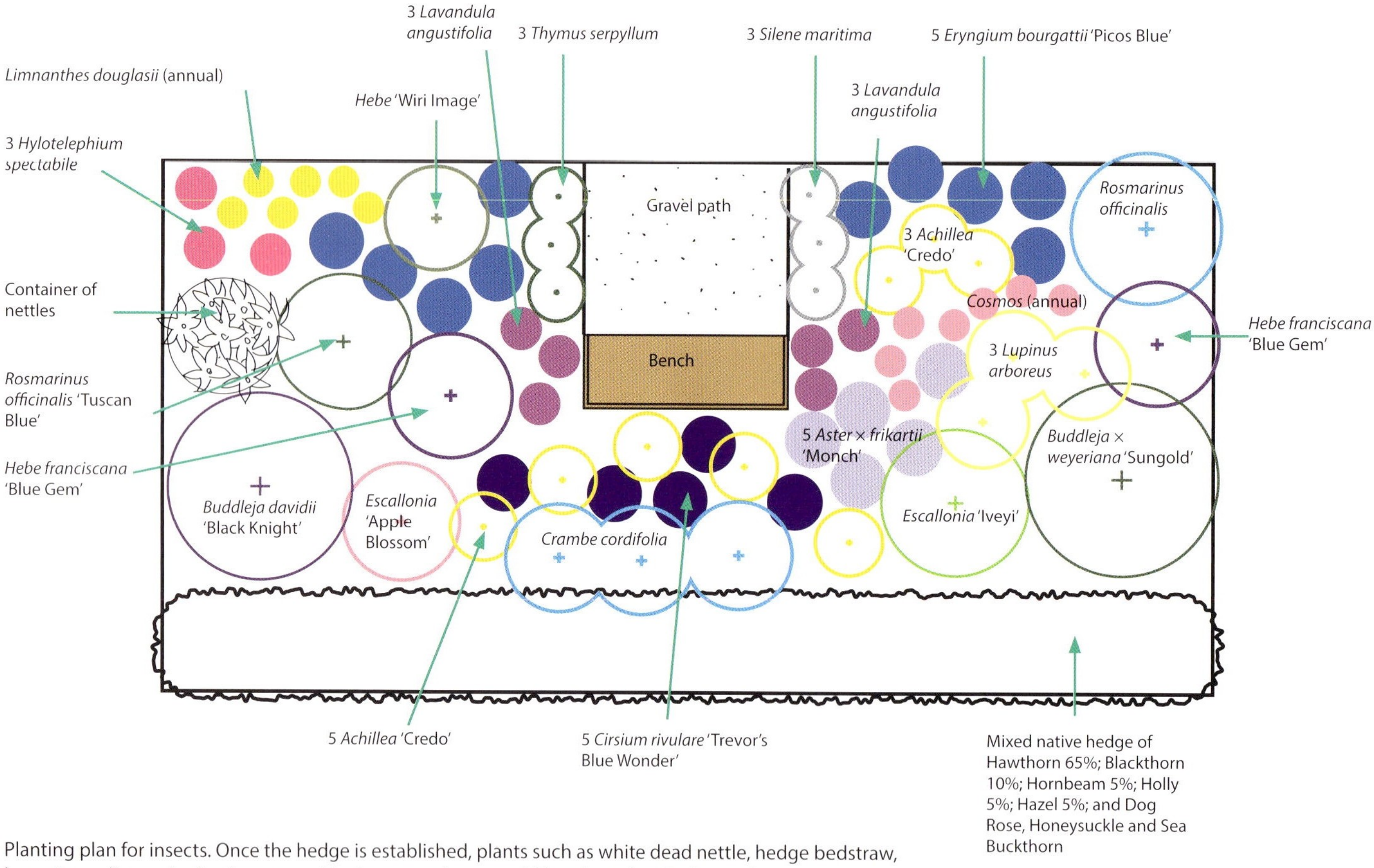

Planting plan for insects. Once the hedge is established, plants such as white dead nettle, hedge bedstraw, honesty, garlic mustard, primrose, and hedge woundwort could be planted at the base of it.

WILDFLOWER LAWNS

A pristine lawn of bowling green quality is exceedingly difficult to achieve on sites that are exposed to the worst of coastal winds. We consider that in many cases it is not worth the effort. If the site is more sheltered from the salt winds, then it is possible but very heavy on energy in terms of mowing it twice a week, scarifying it at least annually, and in hollow tining it and top dressing it. It is heavily reliant on fuel, herbicides, fungicides, and insecticides, and there is a likely loss of some of these chemicals into the ground water within the garden. The lawn will also require regular watering in hot spells of weather.

Early in the year, lesser celandine (*Ranunculus ficaria*) is a major feature in the lawn.

The product of all this is a monoculture with extremely low biodiversity. In the not-too-distant past the advice to gardeners was to kill worms in the lawn using powdered lead arsenate! At least those days have gone. This applies to all gardens, coastal or inland, but coastal gardens will also require a sheltered area away from the worst of the coastal conditions. Overall, if we are interested in achieving as wide a biodiversity as is possible, we should realise that the traditional bowling green lawn, so loved by previous generations of gardeners, is looking increasingly out of date, not only for its environmentally unfriendly reliance on high input of water and chemicals but also for its rather boring monoculture of plant varieties. We can do better.

One option is to use no chemicals and cut the lawn less regularly and not as low as normal. The plants that thrive in lawns have evolved to thrive in grass despite some mowing or grazing and some wildflowers should appear naturally. If this does not work, or a greater diversity of wildflowers is desired, an alternative approach is to plant wildflowers into the existing lawn. Overseeding an existing lawn with wildflower seed does not usually work and our advice is to use wildflower plugs and plant them when the soil is moist. Then keep well-watered until the plugs are established. Depending upon the choice of plants, different parts of the lawn may require cutting at different times. Mowing may be delayed for parts of the lawn that are in flower. Such areas can be further away from the house, confined to edges of the lawn or perhaps a circular area within the lawn. Allowing plants to set seed or not is a way to control the increase in the number of plants such as cowslips (*Primula veris*) or plantains (*Plantago* cultivars). Red clover (*Trifolium pratense*) will rely totally on seeding. They will seed into the more formal planting in borders but should be easily weeded out, or the lawn can be cut before seed is set. Some thought should be given concerning the choice of double flowers. Double flowers will deprive insects of pollen and nectar, but a few chosen double-flowered cultivars will do no harm and will add interest to the overall planting. What we are trying to achieve is for the lawn to be home to a greater variety of small creatures and a food source for pollinating insects and birds.

Plants for a wildflower lawn in partial shade that does not become bone dry

Bugle (*Ajuga reptans*) – if flowers are required then the lawn should not be cut too low. The flowers are usually blue but there are white and pink ones. Many cultivars are in cultivation. *A. reptans* 'Braunherz' is a reliable one with rich purple leaves. *A. reptans* 'Page's Yellow' has attractive yellow foliage, particularly in spring and early summer. It makes a good contrast to *A.* 'Braunhertz'. Bugle is a good source of nectar for several insects including our two bee hawk-moths. Both hawk-moths resemble bumblebees but are more agile. They are daytime flyers. The food plants for the larvae of the narrow-bordered hawk-moth (*Hemaris tityus*) are various scabious species and for the broad-bordered hawk-moth (*Hemaris fuciformis*) honeysuckle (*Lonicera periclymenum*).

Creeping buttercup (*Ranunculus repens*) – this buttercup can be a menace, spreading rapidly across borders, but is likely to be less of a problem in lawns. It may not be popular with many gardeners, but the flowers are visited by hoverflies, bees, butterflies, and wasps in search of pollen and nectar. *R. repens* 'Buttered Popcorn' is a scarce cultivar with a lot of yellow in the foliage but the colour is best in full sun.

These two cultivars of *Ajuga reptans*, 'Braunhertz' and 'Page's Yellow' will add interest to a lawn in partial shade.

Daisies can make an attractive feature in the lawn. Named cultivars should perhaps be planted in the border or in containers and given more attention.

Fritillaria meleagris, the snake's-head fritillary, is native to the United Kingdom and unlike most fritillaries requires a damp spot in the garden.

The dog violet, *Viola riviniana*, requires partial shade in order to thrive. It will increase by self-seeding.

Daisy (*Bellis perennis*) – this daisy prefers cooler and damper conditions to do well. It is pollinated by bees, flies, and beetles. There are named cultivars which are worth some consideration. *B.* 'Miss Mason' is an attractive double white. It requires frequent division and replanting. *B.* 'Dresden China' is another double and is a beautiful rose-pink. It appears to struggle in lawns and requires a lot of care.

Cuckoo flower (*Cardamine pratensis*) – this is a beautiful wildflower which requires some moisture in the soil. If the lawn has a damp spot, then do not mow this area until flowering has finished. The pale pink flowers appear in late spring. It is a food plant for the larvae of the orange tip and green-veined white butterflies. There is a commonly available double-flowered form in cultivation which will act as the food plant. It does not set seed but if cut pieces are left on the lawn after mowing, they will root.

Snake's head fritillary (*Fritillaria meleagris*) – another beautiful plant which requires similar conditions to the cuckoo flower. If a moist spot is available, it will thrive in full sun. Plant the bulbs at a depth of four times the size of the bulb. It may be attacked by the lily beetle whose larvae cover themselves in excrement. It should be allowed to self-seed.

Ragged robin (*Lychnis flos-cuculi*) – ragged robin will thrive in wet areas and is a common plant of damp meadows and has pink or white flowers. It flowers from May to July and can be allowed to self-seed before that area of the lawn is cut. It will attract bees, honeybees, and butterflies. It is a food plant of the lychnis moth (*Hadena bicruris*) which is commonly found in gardens.

Dog violet (*Viola riviniana*) – often found on lawns except on acid soils. The flowers, which are unscented, appear from April to June, and a good-sized patch of it may attract the silver-washed fritillary (*Argynnis paphia*) and the small pearl-bordered fritillary (*Boloria selene*) for which it is the larval food plant.

Germander speedwell (*Veronica chamaedrys*) – a creeping perennial which will form decent sized patches in the lawn. It flowers from March to July and is a good nectar source for solitary bees. The flowers are a beautiful blue colour. It also performs well at the base of hedges.

Ground ivy (*Glechoma hederacea*) – a creeping perennial which provides excellent ground cover. Lilac flowers appear from March to May and are a good nectar source.

Lesser celandine (*Ranunculus ficaria*) – a very invasive plant in borders but that is not a problem in the lawn. It drew the attention of Wordsworth who wrote a poem about it. It is in leaf from December and has disappeared below ground by the end of May. It attracts flies, honeybees, and pollinating beetles. In the United Kingdom there is a national collection of its cultivars. *R. ficaria* 'Salmon's White' is worth a place in the lawn, as is *R. ficaria* 'Brazen Hussy' which has foliage that is almost black.

Oxlip (*Primula elatior*) – the oxlip grows best in shade and does well on clay soils. It flowers from April to May and is very effective when grown with germander speedwell, but do keep it separated from the primrose. It is a source of nectar for early bees and butterflies which will also pollinate it.

Primrose (*Primula vulgaris*) – the primrose is one of the United Kingdom's favourite flowers and looks good when in association with bugle and cuckoo flower. There are many different coloured cultivars in horticulture, and it is a matter of choice as to whether these are planted in the lawn. Personally, we would not choose to do so. It will hybridise with the oxlip (*Primula elatior*) and the cowslip (*Primula veris*) if these are growing in the vicinity. The primrose is a particularly good source of nectar, and the dark-edged bee fly (*Bombylius major*) will be seen feeding using its long, straight proboscis. It is a food plant for the least yellow underwing (*Noctua interjecta*), a moth which is found mainly in southern England where it is widespread but not common.

Selfheal (*Prunella vulgaris*) – selfheal is likely to appear in the lawn once the amount of mowing is reduced. It has red-purple flowers and will flower all summer. If the lawn is fairly regularly mowed, it will form a dense carpet, but it can cope in long grass as well. It is a good nectar plant and attractive to bees.

Ranunculus ficaria 'Brazen Hussy' will add more interest to the lawn. It will hybridise with the normal green-leaved form and a number of the resulting seedlings will have attractively patterned foliage.

Ranunculus ficaria 'Picton's Double' is an attractive double-flowered form of the lesser celandine. It is better to plant this, and similar cultivars, in a shady border rather than the lawn.

The primrose is such a good plant. It likes partial shade and performs well in borders, lawns and at the base of hedges. It is particularly good when planted at the top of grassy banks and will self-seed down the bank.

Primula 'Miss Indigo' is a beautiful double-flowered cultivar of the primrose. Perhaps it should not be planted in natural situations such as lawns and grassy banks where it will look out of place.

Plants for a well-drained and sunny wildflower lawn

Bird's foot trefoil (*Lotus corniculatus*) – once established, this attractive perennial is very drought resistant due to its long tap root. It is pollinated by bees who feed on nectar as do butterflies. It flowers from late June through to autumn and is a food plant for the common blue butterfly (*Polymmatus icarus*) and the burnet moths (*Zygaena* spp.). The plant is common throughout the United Kingdom, often occurring along the coast. It is a low-growing plant and will perform best if the grass is kept relatively short. It is easily grown from seed and if the seed is sown in the autumn, then it will probably flower in its first summer.

Autumn hawkbit (*Leontodon autumnalis*) – a good late nectar plant flowering from July to October. It is pollinated by a wide range of insects.

Cat's ear (*Hypochoeris radicata*) – this perennial is similar to the autumn hawkbit, both of which have flowers that resemble those of the dandelion. It is an excellent nectar plant and copes well with drought conditions. It is a food plant for the feathered footman (*Coscinia striata*). This attractive moth is quite common in Europe but unfortunately very rare in the United Kingdom. Perhaps it will become more common as a result of global warming.

Cowslip (*Primula veris*) – this is a primula for sunny places especially on alkaline soils. It will tolerate normal mowing from late spring to autumn. It is pollinated by bees and hoverflies.

Dandelion (*Taraxacum officinale*) – they mostly undergo apomictic reproduction in which fertile seed is set without sexual cross-pollination. As a result, they breed true and there may be several different clones within the garden. If you are concerned about an excess of seedlings the flower heads should be removed before the seed ripens. They are an early source of nectar for bees and butterflies. There is a white-flowered species of dandelion, *Taraxacum albidum*, which is less robust than our native species but could be worth experimenting with for something different. It flowers from April onwards.

Greater plantain (*Plantago major*) – modern gardeners are not impressed by the greater plantain and view it as a weed but for the Anglo Saxons it was one of nine sacred herbs and believed to have medicinal and magical properties. It is the food plant for several moths including rustic (*Hoplodrina blanda*), the very similar uncertain (*Hoplodrina octogenaria*), and treble lines (*Charanyca trigrammica*). These three moths are common throughout England and Wales and likely to come into the garden. There is a cultivar available, *P. major* 'Purple Perversion' that has conspicuously fringed purple-red leaves and is a vigorous grower. *P. major* 'Rosularis' is an old cultivar which produces a rosette of green leaves rather than the 'rat's tail' flower spike. It would possibly object to losing its head to regular mowing so it may be better to grow it in the border. Both these cultivars will self-seed.

The plantain 'Purple Perversion' will add colour to the lawn but may suffer from mildew in very dry weather.

Taraxacum albidum is easily grown from seed and the resulting seedlings can then be experimented with in the lawn.

The rose plantain (*Plantago major* 'Rosularis') is an antique plant and deserves greater appreciation.

Hoary plantain (*Plantago media*) – this occurs in the southern half of England but not on acidic soils. It has the most attractive flowers of all the plantains. It is pollinated by wind or insects, particularly bees.

Ribwort plantain (*Plantago lanceolata*) – this attracts hoverflies and is the food plant for several species of moth and the Glanville fritillary (*Melitaea cinxia*), a butterfly which is confined to coastal landslips on the Isle of Wight, the Channel Islands, and a few locations on the south coast of the United Kingdom. There is an interesting 'white whiskered' cultivar. The flowers on the spike have unusually long stigmas which give the spike the appearance of a bottle brush.

Meadow buttercup (*Ranunculus acris*) – grows well on alkaline soils and flowers from May to July. It will repeat flower if cut back. It will need to be in an area of the lawn which is not mown until after flowering. It is rich in nectar and a food plant for the flame brocade (*Trigonophora flammea*). This moth is resident in the Channel Islands and is an occasional immigrant to the south coast of England as far north as Hampshire. It is on the wing in October and November. Other food plants are lesser celandine, broom, and blackthorn, all of which we have suggested could be planted in the coastal garden.

Mouse-ear hawkweed (*Pilosella officinarum*) – a low-growing perennial with yellow flowers from May until autumn. The flowers are rich in both pollen and nectar. It looks attractive when grown with dwarf thymes.

Red clover (*Trifolium pratense*) – this clover can still be in flower in November and is renowned as being an excellent nectar source for bumblebees. It can be short-lived and does not like becoming very dry. It is an excellent food plant for various moths such as the marbled clover (*Heliothis viriplaca*). This moth inhabits grassland and sand or shingle beaches. It is not particularly common in the south of England but even rarer elsewhere.

White clover (*Trifolium repens*) – this is a creeping perennial that performs well on clay soils. It struggles in tall grass but thrives in closely mown turf. It flowers all summer and remains green when the grass is brown. It is a must for bumblebees. There are a number of cultivars that are worth trying in the lawn. *Trifolium repens* 'Purpurascens Quadrifolium' has purple leaves edged in green and will have a number of stems bearing four leaves. *T. repens* 'Green Ice' is very vigorous with two-toned green leaves. *T. repens* 'William' has foliage that changes colour with the seasons. It has magenta-coloured flowers.

Slender speedwell (*Veronica filiformis*) – this is a non-native species which was introduced into the UK during the early nineteenth century. It has become naturalised, especially in lawns, and flowers in April producing a carpet of blue. Cut pieces of stem root readily if left on the lawn after mowing.

Orchids – orchid seed travels long distances and several orchid species could grow in the lawn. Plants may not reach flowering size for several years and therefore the gardener needs to recognise the young plants by their leaves. Orchids that may appear in the lawn include bee orchid (*Ophrys apifera*), twayblade (*Neottia ovata*), and green winged orchid (*Anacamptis morio*). Autumn ladies' tresses (*Spiranthes spiralis*) is a small orchid that often occurs in lawns along the coast from South Devon to North Yorkshire. It grows to about 4 inches tall and the flowers are very fragrant. The gardener should recognise the leaves before they disappear by the end of June. The flower stems will appear in the autumn along with a new rosette of leaves. Allow to seed and then mowing can restart.

Trifolium repens 'Green Ice'.

Both 'Green Ice' and 'William' are cultivars of white clover (*Trifolium repens*) and have attractive coloured foliage which will add interest to the lawn.

COMPOST HEAP

Every garden, large or small, coastal or inland, should have a compost heap. We would recommend, if there is sufficient space, that the garden should have three. One can be in the process of suitable material being added to it to rot down; one is full and in the process of producing quality compost; and the third is well rotted and being used in the garden. A very economical way of making a compost heap is to use pallets held in place by secure stakes. The front pallet can be held in place by wire which allows easy removal to get at the compost when it requires mixing or removal. An alternative is to purchase plastic compost bins which are perfectly adequate but make it less easy for larger wildlife to get in and out of the decaying compost. As well as providing compost for the garden the heaps will be home to a large range of invertebrates and larger animals, e.g. grass snakes, hedgehogs, and amphibians.

Adding compost to your garden soil on a regular basis will improve its fertility, improve the drainage of clay soils, and improve the water retention of sandy ones. The secret to achieving good compost is to have a balance between nitrogen-rich materials (the 'greens') and carbon-rich materials (the 'browns'). A lot is written about the relative quantities of each of these but, to avoid losing any sleep over it, equal amounts of greens and browns will produce quality compost so long as they are kept well mixed. The 'greens' will include grass cuttings, soft green prunings, weeds, raw vegetable and fruit peelings and tea bags. Weeds that have gone to seed and perennial weeds such as ground elder and bindweed should be avoided. The leaves of comfrey and nettles (not roots) rot down quickly and increase the temperature of the heap. The temperature of the average garden compost heap will not be high enough to kill all weed seeds.

The 'browns' will include cardboard, egg boxes, wastepaper, hay and straw, fallen leaves and woody prunings. If bulky items are shredded before addition to the heap, they will rot down more quickly. A possible problem with shredded paper is that it can become compressed and not decay properly. Other items to add to the heap include crushed eggshells, wool, hair, and wood ash. Do not add cooked food, dog and cat faeces and the perennial weeds mentioned earlier. It is important to avoid adding too large a quantity of one material at any one time. It is a common mistake to add too many grass cuttings if a large lawn has been mown. They become compressed within the heap and do not undergo decay due to a lack of oxygen. A 4-inch-thick layer is sufficient, followed by a layer of browns. Every couple of months give the heap a good mixing by turning it over using a garden fork. Heaps that have not been turned will still produce good compost, but not as quickly. There is an argument that not turning the heap is better for the wildlife that is in it. Garden centres will sell compost accelerators; urine diluted to 1 in 20 with water is a cheaper option.

Gardeners in coastal areas may have easy access to seaweed. Rules concerning its removal from beaches vary from country to country. In the United Kingdom you can usually collect seaweed if it is both floating in the water and unattached to rocks etc. If the seaweed is growing on the foreshore or seabed or has been deposited on the shore by the tide then you must obtain permission from the landowner. It is safer to check on local regulations to avoid any potential problem. When the seaweed has been collected, we recommend that it is rinsed with water to wash out excess salt. There is some argument over whether this is necessary, but we feel that rinsing the seaweed will avoid any build-up of salt over time. The seaweed can then be chopped up and added to the compost heap. This will speed up the composting process. If a lot of seaweed is being added, then it is wise to mix it with some woody or fibrous material, i.e. greens plus browns. Seaweed is an excellent addition to the compost heap. It is rich in nitrogen, phosphorus, potassium and magnesium, as well as small quantities of other minerals.

Some care needs to be taken over substances that may produce problems in our compost heap. Hay and straw are useful additions to the heap but may have residues of herbicide and/or pesticide in them. It pays to check where it has come from. Some gardeners will advise not to add leaves of walnut and eucalyptus to the heap as these trees secrete toxins into the soil and therefore several plants struggle under these trees. The evidence is that these toxins will be broken down during the decomposition process and therefore can be added to the heap; just do not add too many. Experiments have shown that a chemical in the peel of citrus fruit inhibits the germination of seeds. It can be added to the compost heap, as the chemical will be broken down within the heap, but, again, do not add huge quantities.

There are lots of micro-organisms within the heap such as bacteria, nematodes, and fungi which feed on the organic material there. These are microscopic and will not attract our attention but there is a host of invertebrates that will. There will be several species of woodlice present. They feed on rotting materials and other decaying vegetation. A healthy compost heap needs worms. Their tunnelling aerates the compost and breaks down the organic matter within the heap. The worm which is dark red or brown with clear orange bands and is common in compost heaps is the brandling (*Elsenia fetida*). This worm is rarely found in soil but thrives in rotting vegetation and compost. The brandling's main predators include blackbirds, robins, starlings and slow worms.

Centipedes (*Lithobius* spp. *et al.*) will be found in the top layers of the heap. They feed upon other invertebrates that are in the heap. Millipedes (*Polydesmus* spp. *et al.*) will also be present. They feed on the decaying vegetation turning it into compost. They also feed upon insect carcases that they come across. The earwig (*Forficula auricularia*) is another common inhabitant of the

Brandling worms (*Eisenia fetida*) are smaller than earthworms and are commonly found in compost heaps. The slug in the picture is a leopard slug (*Limax maximus*) which is commonly found in the top layer of the compost heap and plays an important part in the composting process. It is the gardener's friend.

The earwig (*Forficula auricularia*) is a common resident in the garden. This is a male as the *cerci* are curved; they are straight in the female.

garden including the compost heap. It is active at night and is omnivorous in its choice of food. It may damage our flowers such as dahlias and chrysanthemums but is beneficial in the compost heap, eating both decaying material and a range of small insects. It is easily recognised by the pincers (known as *cerci*) at the end of the body.

The compost heap is an ideal place for slugs. They break down the material in the heap and play an important part in the formation of quality compost. They will lay their eggs within the heap but many of these will be eaten by beetles and other organisms within the heap and will not be a problem when the compost is added to the garden soil. On no account should slug pellets be added to the heap as they will kill many beneficial organisms as well as the slugs. There is no need to kill the slugs, in fact some gardeners add slugs to the heap where they are doing good work. Most slugs prefer to eat decaying leaves and parts of plants that humans do not consume. The tidy gardener suffers more than the untidy one as a lack of dead vegetation drives slugs to eat better parts of the plant. It is a similar argument concerning snails. They will be useful in the compost heap despite the problem of them feeding on living vegetation, and not all species of snail do this. Only two of our snails, the garden snail (*Cornu aspersum*) and the strawberry snail (*Trochulus striolatus*) really do much damage in the garden.

Many species of fly are associated with decaying plant material and will be present in the compost heap. They are not a problem with an open heap but if the gardener is using a compost bin, when the lid is removed they are disturbed and take flight. The flies are not harmful but if their flying is disconcerting then walk away and return a few minutes later when they will have settled back in the bin. Fresh material can now be added. The larvae of the flies play a vital part in the decay process.

The slow worm (*Anguis fragilis*) is a harmless, legless lizard and is also the gardener's friend as it feeds on slugs and snails. It can live for over twenty years.

The compost heap may be a temporary home for larger creatures. Hopefully, rats will not be a problem. It is important not to add cooked food waste to the heap as this will only encourage them. Amphibians such as frogs, toads and newts may visit and even use the heap for hibernating during the winter. Slow worms (*Anguis fragilis*) like a diet of slugs, worms and snails and may visit the heap for easy pickings. If you discover that they are breeding in the heap, then do not disturb them but use another heap. If the heap is being turned, be careful. Turning the heap is not essential, but if done, it is best to do it in May after the slow worms have mated and/or in October when the young have been born. The slow worm is not a worm nor a snake but is a legless lizard. It hibernates from October to March and is totally harmless. It is predated on by the domestic cat.

A piece of corrugated metal placed in a sunny spot provides a warm place for reptiles.

Grass snakes (*Natrix helvetica*) may also breed in the compost heap. Between ten and forty white eggs are laid in June and July and hatch in about ten weeks. If you come across them then, as with the slow worm, do not disturb them but move to another heap. Remember that the grass snake is totally harmless but may feed on some of the frogs within the garden.

Reptiles, but especially slow worms and grass snakes, can be encouraged into the garden by placing a piece of corrugated metal in a low-cut part of the wildflower meadow. It should be put in place by the end of the winter ready for reptiles to emerge after hibernation. Check under it no more than once a week. Move quietly as lizards may be sitting on the top of the sheet. Be careful if the garden is close to heathland or rough grassland as adders (*Vipera berus*) may use it. They are the United Kingdom's only venomous reptile, having a poisonous bite.

Hedgehogs (*Erinaceus europaeus*) are likely to visit the compost heap particularly at night. They will feed on slugs, snails, and other invertebrates. Leave them to it and do not be tempted to put out bread and milk for them. However, when turning the heap, if you must, do look out for hedgehogs that may be at the bottom of it. Similarly, when lighting a bonfire, check underneath it that no hedgehog is present.

WATER CONSERVATION

The ideal garden soil is almost a contradiction in terms. The target to aim for is to have it 'moist and well drained'. Sandy soils drain quickly but do not retain moisture well. On the other hand, clay soils hold onto moisture but often become very wet to the extent that walking on them simply leads to compaction which is not good for the structure of the soil. Particles of clay absorb minerals onto their surface and consequently can be very fertile. Sandy soils contain little clay and hence can become mineral deficient, particularly in wet seasons. Water requires a lot of energy to increase its temperature and hence clay soils are slow to warm up in the spring, but sandy soils will warm up more quickly. There is a balance to aim for, to achieve the ideal conditions. The addition of large amounts of garden compost will increase the water retention of sandy soils but it needs to be added every year; this is not an instant fix. Similarly, the addition of large amounts of compost to the clay soil will improve the drainage by separating clay particles. The addition of horticultural grit is also of benefit. Again, these additions need to be done annually.

Water is a commodity that must be paid for. If global warming continues, as is forecast, then gardeners ought to do what they can to conserve water and use it responsibly in their gardens. As well as planting well and appropriately and using drought-resistant plant species where needed, the following measures can help:

- The use of permeable paving will help to prevent run-off. Rainwater will go into the water table rather than into drains and roadways which can lead to flooding.
- Water butts can be connected to down pipes to collect rainwater. The contents of a 90-litre water butt do not last long, so the water butt should be as large as is possible and more than one butt used. In periods of heavy rain, the butts will fill up quickly. Consideration needs to be given as to where overflow goes.
- Grey water can be used to water the garden. Grey water comes from baths, washing up and similar household sources. However, excess addition of detergents to the soil is not recommended.
- Do not water the lawn. Allow the grass to turn brown during spells of dry weather; it will recover when the rain returns. An alternative is to rethink the amount of lawn that is required (*see* above, page 158).
- Do not use sprinklers to water the garden. They use a lot of water and much of it is wasted. When watering, add the water to where it is needed, i.e. to the base of the plant. Established trees and shrubs should be able to look after themselves. If not, then it is the wrong plant for that situation. When watering, add a good amount every few days and not a little every day. The aim is to avoid attracting feeder roots to the surface.
- Containers may require watering every day during periods of hot weather. The use of drought-resistant plants, e.g. pelargoniums and succulents, will reduce the amount of water required.
- Rainwater harvesting systems can be installed by specialist contractors. These systems can supply the house to flush lavatories and for use in washing machines, as well as to irrigate the garden.

In late winter, a thick layer of mulch should be added to as many planting areas as is possible. This does not include areas such as shingle beds, wildflower meadows, and areas containing a lot of drought-resistant plants such as lavender, santolina and yucca which should look after themselves. Garden compost makes a particularly good mulch, as does leaf mould. Chopped bark can be used but as it decomposes it will reduce the amount of nitrogen in the soil. The same is true of lawn mowings. Spent mushroom compost, when available, makes a good mulch but is alkaline, so do not use it around ericaceous plants such as rhododendrons, pieris and camellias. Gravel or shingle can make an attractive mulch but does get hot in the summer. It is best used in combination with drought-resistant planting. When mulching, do not apply it so that it rests against the stem or trunk of the plant as it may introduce rot. If the layer is very thick on top of dormant bulbs, they may be unable to grow through it. A layer that is 5 centimetres thick is our recommendation.

DROUGHT-RESISTANT PLANTS

Many drought-resistant plants have silver or grey leaves which reflect the rays from the sun. Some have coatings of fine hairs on their leaves helping to increase humidity around the plant tissue. A selection of suitable plants is listed in Table 6.1. More information on these plants can be found in Chapter 5.

TABLE 6.1 SUGGESTED DROUGHT-RESISTANT PLANTS

Trees and shrubs

- *Agave americana*
- *Artemisia arborescens* 'Powis Castle'
- *Brachyglottis* 'Sunshine' and *Brachyglottis* 'Drysdale'
- *Buddleja davidii* and *Buddleja weyeriana* 'Golden Glow'
- *Ceanothus* cultivars
- *Chamaerops humilis*
- *Cistus* cultivars
- *Convolvulus cneorum*
- *Elaeagnus angustifolia*
- *Fascicularia bicolor*
- *Lavandula* cultivars
- *Phlomis fruticosa*
- *Phormium* cultivars
- *Rosmarinus officinalis*
- *Santolina* cultivars
- *Spartium junceum*
- *Trachycarpus fortunei*
- *Ulex europaeus*
- *Yucca* cultivars

Perennials, grasses and climbers

- *Achillea* cultivars
- *Agapanthus* cultivars
- *Allium* cultivars
- *Calamagrostis* cultivars
- *Campsis radicans*
- *Campsis* × *tagliabuana* (climber)
- *Centranthus ruber*
- *Crambe maritima*
- *Eccremocarpus scaber* (climber)
- *Echinops ritro*
- *Elymus magellanica*
- *Eryngium* cultivars
- *Hylotelephium* cultivars (formerly *Sedum*)
- *Iris germanica*
- *Kniphofia* cultivars
- *Panicum virgatum* cultivars
- *Sisyrinchium striatum*
- *Stipa tenuissima*
- *Thymus serpyllum*

Annuals and perennials grown as annuals

- *Arctotis* cultivars
- *Eschscholzia* cultivars
- *Gazania* cultivars
- *Mesembryanthemum criniflorum*
- *Osteospermum* cultivars

Appendix I

Gardens to Visit

UNITED KINGDOM

Abbotsbury Subtropical Gardens, Weymouth, Dorset DT3 4LA
A magnificent garden set in 30 acres of a woodland valley. The valley is well protected from the salt winds and the microclimate enables many rare and unusual species to thrive.
Website: https://abbotsbury-tourism.co.uk/gardens

Arundell Garden, West Pentire, Crantock, Cornwall TR8 5SE
On an exposed headland between two beaches, this is a 1-acre garden with features including a gravel garden, exotic garden, pond and herbaceous border. It has wheelchair access and toilets.
Email: fundraising@cornwallhospice.co.uk

Chygurno, Lamorna, Penzance, Cornwall TR19 6XH
A 3-acre clifftop garden with steep paths. Many South African and New Zealand native plants thrive here. Limited opening times.
Telephone: 01736 732153

Compton Acres, 164 Canford Cliffs Road, Poole, Dorset BH13 7ES
Compton Acres is home to a magnificent collection of trees, shrubs and herbaceous perennials, some of which grow nowhere else in the country. It is a fine example illustrating the effect of windbreaks on plant growth in sheltered areas.
Website: www.comptonacres.co.uk

Driftwood, 4, Marine Drive, Seaford BN25 2RS
A prize-winning garden described by Monty Don as 'a small garden by the sea with inspired planting and design.'
Website: www.driftwoodbysea.co.uk

2 Durnamuck, Little Loch Broom, Wester Ross IV23 2QZ
A southeast-facing coastal garden with a rich mix of perennials, trees, shrubs and vegetables.
Email: sueandwill@icloud.com
Facebook: '2 Durnamuck'

The Gertrude Jekyll Garden, Holy Island, Berwick-upon-Tweed, Northumberland TD15 2SH
A small walled garden in a spectacular location featuring a combination of hardy annuals, perennials and vegetables. The garden is open daily, but tide times vary.
Telephone: 01289 389244
Email: lindisfarne@nationaltrust.org.uk

Old Vicarage Garden, East Ruston, Norwich, Norfolk NR12 9HN
This garden is in an exposed landscape 1.5 miles from the North Sea. It has large shelter belts and includes an exotic garden, Mediterranean garden, wildlife garden, wildflower meadow and a vegetable garden, amongst others.
Website: www.eastrustonoldvicarage.co.uk
Email: erovoffice@btconnect.com

Prospect Cottage, Dungeness Road, Romney Marsh TN29 9NE
The late Derek Jarman's home and garden illustrates what can be achieved in the most exposed of coastal sites. It can be seen from the road.

The Lookout, Kilmuir, North Kessock, Inverness IV1 3ZG
A three-quarter-acre garden which is only for the sure footed. Designed to encourage wildlife and features a sheltered courtyard, productive area, a scree and rockery area and a pond. By appointment only.
Email: david@veitch.biz

The Secret Seaside Gardens of Blackpool, Blackpool Sands, Stoke Fleming, Dartmouth TQ6 0RG
A recently restored nineteenth-century plantsman's garden surrounded by a shelter belt of evergreens and pines. Many specimens from the southern hemisphere thrive in this coastal location. Open daily April to September (subject to weather).
Website: www.blackpoolsands.co.uk
Telephone: 01803 771801

Tresco Abbey Garden, Tresco, Isles of Scilly TR24 0QQ
This large garden is home to 20,000 plants from eighty countries. Refer to the website (www.tresco.co.uk) for details on travelling there. The gardens are open daily all year round.
Email: gardenvisitorcentre@tresco.co.uk

Ventnor Botanic Garden, Undercliff Drive, Ventnor, Isle of Wight PO38 1UL
A 22-acre garden in a sheltered and warm microclimate. Open daily, it is home to 6,000 species of plants selected to cope with the shallow alkaline soil and the salt winds.
Website: www.botanic.co.uk
Email: info@botanic.co.uk

REST OF THE WORLD

Coastal Maine Botanical Gardens, 105 Botanical Gardens Drive, Boothbay, ME 04537, USA
The gardens cover 300 acres and have many areas of interest including a meditation garden and a shoreland trail featuring indigenous plants.
Website: www.mainegardens.org
Email: info@mainegardens.org

Giardini La Mortella, Via Francesco Calise, 45, 80075, Forio (NA), Italy
This sub-tropical and Mediterranean garden on the Island of Ischia was the home of William Walton and his wife Susana. It has been developed since 1956.
Website: www.lamortella.org/eng
Email: info@lamortella.org

Harold Porter National Botanic Garden, Cnr Clarence Drive & Broadwith Road, Betty's Bay, Western Cape, South Africa
The garden is open daily and is situated in the centre of the coastal fynbos where the flora is at its richest. Renowned for its waterfalls, pools and its wide-ranging biodiversity.
Website: www.sanbi.org/gardens/harold-porter
Email: Harold.Porter.NBG@sanbi.org.za

Marimurtra Botanical Garden, Passeig de Carles Faust, 9, Blanes (Girona), Spain.
Founded in 1924, this is one of the most beautiful gardens in Spain. It is at the top of steep cliffs and contains around 4,000 species of plant from the five continents. There are splendid views over the sea.
Website: www.marimurtra.cat
Email: marimurtra@marimurtra.cat

Mendocino Coast Botanical Gardens, 18220 North Highway One, Fort Bragg, California CA95437, USA
A garden of 47 acres, fronting the Pacific Ocean, and featuring gardens of perennials, vegetables, and woodland planting. The coastal bluff and marine zone feature plants adapted to poor soils and salt-laden winds. Whales may be spotted from here.
Website: www.gardenbythesea.org
Email: info@gardenbythesea.org

Santa Clotilde Gardens, Lloret de Mar, Girona, Catalonia, Spain
This is a garden on a cliff top with a Renaissance inspired design and stunning views. The planting uses plants typical of the Mediterranean basin. The large staircase in the garden has been described as one of the coastlines most beautiful scenes.
Website: www.lloretdemar.org
Email: central-turisme@lloret.cat

Villa Cimbrone, Via S. Chiara – 84010, Ravello (Sa), Italy
The hotel has a garden which is open to the public. The garden was created with the help of Vita Sackville-West and is laid out along the clifftop with a magnificent seascape. It contains elements of both Italian and English garden design.
Website: www.hotelvillacimbrone.com
Email: info@villacimbrone.it

Villa Ephrussi de Rothschild, 1, Avenue Ephrussi de Rothschild, 06230, Saint-Jean-Cap-Ferrat, France
The gardens are open 365 days a year. The French Garden is the main garden and is designed in the form of the bridge of a boat with the sea on either side and is adorned with waterfalls and ponds. The Spanish Garden contains many fragrant plants. There are six other gardens including The Provençal Garden and The Exotic Garden. The whole is a treat for the senses.
Website: www.villa-ephrussi.com
Email: message@villa-ephrussi.com

Appendix II

Plant Nurseries and Places to See Coastal Plants

Atlantic Botanic, Whitegates, Moor Lane, Braunton, Devon EX33 2NU
A small family-run nursery which specialises in growing plants that thrive on the coast.
Website: www.atlanticbotanic.co.uk
Email: atlanticbotanic@live.co.uk

Burncoose Nurseries, Gwennap, Redruth, Cornwall TR16 6BJ
A large nursery plus a 30-acre woodland garden. Good choice of suitable plants. Mail order available and gardens planned and planted.
Website: www.burncoose.co.uk
Email: info@burncoose.co.uk

Cross Common Nursery, The Lizard, Helston, Cornwall TR12 7PD
A small family-run nursery specialising in tropical, exotic and coastal plants.
Website: www.crosscommonnursery.co.uk
Email: info@crosscommonnursery.co.uk

Garden Cottage Nursery, Tournaig, Poolewe, Achnasheen IV22 2LH
A family-run nursery with a wide selection of plants suitable for windy, wet climates. Has a very informative website.
Website: www.gcnursery.co.uk
Email: ben@gcnursery.co.uk

Knoll Gardens, Stapehill Road, Hampreston, Wimborne BH21 7ND
A nursery and associated garden, specialising in grasses and perennials, many of which are suitable for coastal planting.
Website: www.knollgardens.co.uk
Email: enquiries@knollgardens.co.uk

The Coastal Gardener, Fakenham Farm, Eddington Road, St. Helens, Isle of Wight PO33 1XS
A small specialist plant nursery selling a wide selection of coastal plants.
Website: www.thecoastalgardener.co.uk
Email: thecoastalgardener@icloud.com

Trevena Cross Nurseries, Breage, Helston, Cornwall TR13 9PY
Has a huge collection of coastal plants and grows more than 90 per cent of the plants it sells. Mail order available.
Website: www.trevenacross.co.uk
Email: sales@trevenacross.co.uk

Glossary

Annual Plants that complete their growing cycle in one year and then die.

Apomictic reproduction Asexual reproduction through seed. The dandelion (*Taraxacum officinale*) is a common example.

Batter The receding slope of a wall. The opposite of an overhang.

Biennial Plants that complete their growing cycle over two years and then die.

Biodiversity All the varieties of life, both animal and plant, found in the garden.

Bolting Plants, especially vegetables, prematurely running to seed.

Bug hotel A structure to provide shelter for a range of insects.

CAD Computer aided design.

Calcareous Soils or sediments described as calcareous contain a high proportion of calcium carbonate and so are alkaline. This can be from the remains of marine organisms such as shells or the compressed microscopic plankton that form chalk.

Cloche A low structure, traditionally glass but now usually made from plastic, used to cover plants and provide protection against the weather and pests.

Constancy Bumblebees sometimes concentrate on a group of a single species of flower when foraging.

Coping A stone cap on the top of a freestanding wall to prevent water damage and create a neat finish.

Corm A swollen, underground stem that acts as a storage organ during the winter, e.g. *Crocosmia*.

Cottage garden style A very popular, traditional, and informal style of planting that creates a rich tapestry of colours from flowers and foliage arranged in an uncontrived style. It is likely to be quite high maintenance and wildlife friendly.

Cultivar A particular form or variety of a plant produced in cultivation by selective breeding, e.g. *Crocosmia* 'Limpopo'.

Cuttings Pieces of plant material that are used to asexually propagate a wide range of shrubs and perennials.

Datum point A set reference point against which other measurements, such as levels, are made.

Dead heading The removal of dead flowers from plants to encourage the production of more bloom.

Deciduous Trees and shrubs that shed their leaves on an annual basis, usually in the autumn.

Division A method of propagation for perennials with spreading root stocks where a single plant is separated into several new plants.

Ericaceous Acidic soil or compost, or plants that favour acidic soils and do not tolerate alkaline ones.

Fastigiate A columnar form of a tree or shrub that has a naturally tall and narrow shape with its branches pointing upwards.

Fertiliser Concentrated form of plant nutrients used to enhance growth.

Focal point A striking plant or hard landscaping feature such as a sculpture, carefully placed by the designer for maximum effect.

Gabions Wire baskets or cages filled with rocks and used in garden landscaping for retaining or decorative walls. They are often used in civil engineering as a sea defence so can help create a coastal style in gardens.

Ground cover Low-growing perennial plants or shrubs that form an attractive carpet and supress weeds.

Groyne A rigid structure built out from a beach to prevent the erosion caused by longshore drift. In the United Kingdom they are traditionally made from timber but can also be built from rocks or gabions.

Halophyte A plant that has adapted to cope with high concentrations of salt in soil and wind.

Hard landscaping The built elements of a garden, such as patios, walls or pergolas, that are made from materials such as paving stones or timber, rather than the plants used in soft landscaping.

Herbaceous Plants with non-woody stems which die back totally in the late autumn and winter.

Hollow tining The removal, using a specialised tool, of small plugs from the lawn in order to allow air and water to penetrate. It is usually carried out in autumn.

Inorganic fertiliser A product of the chemical industry consisting of plant nutrient minerals and synthetic chemicals.

Invertebrate Animals without a backbone such as moths, slugs, woodlice and spiders. They play an important part in the food chains in the garden.

Levels The relative heights of points in a garden. These are measured by various means during the survey.

Lichen Lichens are a complex life form consisting of fungi living in a symbiotic partnership with algae. They occur in a very wide range of habitats but are sensitive to air pollution.

Longshore drift The process by which sediment is moved along the coast line by the pressure of currents and waves.

Minimalist garden design A streamlined, uncluttered style with strong clear lines and a limited palette of colours, plants, and materials. Done well, it should have a calm and modern feel and tends to be low maintenance.

Monoculture The growing of one plant species within a given area.

Mulching Covering the ground with organic material to suppress weed growth, retain moisture and improve the structure of the soil.

Narrative In garden design this could refer to the sequence of elements and focal points that are presented as you move around the garden. Attention is drawn to different plants, hard landscaping materials, views, scents, and feelings of space, in a sequence that should, if the garden is well designed, be intentional and carefully considered by the designer. Gardens also change a great deal throughout the year and over the years, and this is also part of the narrative of the garden and an important design consideration.

Native hedge A hedge consisting entirely of native plants.

Native plants Plants that grow naturally within a specific area or country and were not introduced by humans.

Offsets Small plants that form around the base of the parent plant. Can be used for propagation.

Organic fertiliser Organic fertilisers are those derived from animal or plant sources e.g. fish, blood and bone. They are slower acting than inorganic fertilisers.

Perennials Plants that live for more than two years. The term is also often used to refer to plants that have little or no woody growth, in contrast to trees and shrubs.

Pergola An open lattice structure of uprights and crossbeams used to create a shaded path or sitting area. It is traditionally built with timber uprights or brick pillars with timber crossbeams, but increasingly materials such as aluminium or vinyl are used. Climbers are often grown over the structure.

pH A measure of the acidity of the soil. 1–6 is acidic, 7 is neutral and 8–14 is alkaline.

Pleached hedge Consists of trees grown with a straight, clear stem and a head of branches that are trained on a framework to form a narrow, raised hedge.

Plug plants Young plants grown singly in modular trays. They are more affordable than established plants and limit root damage when planting.

Prairie planting style A style of planting using perennials and grasses to create an effect originally inspired by the grasslands of the American Midwest. It tends to use a more limited palette of plants than a cottage garden but with larger quantities of each plant, planted in naturalistic drifts or blocks. When designing a prairie-style planting scheme, the advice is to 'halve the species but double the quantities'.

Productive area Part of the garden designed for the growing of fruit and vegetables.

Pruning The trimming of plants to remove unwanted or dead growth. It is used to control the growth, size and flowering of plants.

Reversion Usually refers to the loss of variegation in plants, with it being replaced by normal, single colour foliage.

Scarification The mechanical removal of moss, dead grass and other material from a lawn.

Scorch The development of brown or dead areas of foliage. Usually caused by excessive sun, wind or, in coastal areas, salt.

Screening The use of trees, shrubs, trellis and other materials to hide unwanted views from the garden.

Shelter belt A more substantial planting than a windbreak, used to create a sheltered area within the garden.

Soft landscaping The planted elements of a garden, such as trees, shrubs and perennials, as opposed to the building materials used in hard landscaping.

Suckers Shoots that grow from the root systems of various species of trees and shrubs. They are often considered an unsightly nuisance but can be used for propagation.

Tap root A central, dominant root of a plant, growing vertically downwards.

Top dressing The spreading of a thin layer of compost over the surface of the lawn.

Topsoil The upper layer of soil, typically consisting of a mixture of sand and grit, clay and decomposed organic matter.

Transpiration The loss of water through the leaves of a plant.

Tree Preservation Order (TPO) In the United Kingdom, trees with a TPO cannot have work done on them without permission from the local planning authority.

Variegation The appearance of differently coloured areas within leaves.

Windbreak A large shrub, hedge or staggered rows of trees and shrubs that filter the wind reducing its effect on the leeward side of the planting.

Index

An entry in **bold** indicates a photograph.